Why Japan Lost World War II

James Biser Whisker
and
John R. Coe

Why Japan Lost World War II

James Biser Whisker
and
John R. Coe

Academica Press
Washington~London

Library of Congress Cataloging-in-Publication Data

Names: Whisker, James B. author. | Coe, John R. author
Title: Why japan lost world war ii / James B. Whisker and John R. Coe
Description: Washington : Academica Press, 2020. | Includes references.
Identifiers: LCCN 2020946302 | ISBN 9781680539479 (hardcover)

Dedicated to
Donald Coe, Sr. DDS
and
James Edwin Whisker
veterans of the Pacific War

John R. Coe was born and raised in Parkersburg, WV. He was the youngest child of the late Dr. Donald E. Coe and Jacqueline Ann Galka. He received his B.S. degree in biology from Glenville State College, his MS from WVU, and his Ed.D. from Liberty University. He is also a retired central office administrator from the Wood County schools. John is currently an independent businessman. The businesses include an analytical chemistry lab as well as a restaurant. He is an avid firearm collector of pistols, rifles, and muzzle loaders. He can be found almost any day on the rifle range. He has been married 31 years to his college sweetheart, Rosemary.

James B. Whisker is professor emeritus from West Virginia University, where he taught for over 37 years. He was adviser to the WVU College Republicans during most of his tenure. He received his Ph.D. from the University of Maryland in 1969. He is author or co-author of books on the gunsmiths and arms makers of eighteen states. He co-authored books on clockmakers and silversmiths of Pennsylvania and Maryland. Among his other books are *The Alien Tort Claims Act, The Militia, The Right to Hunt; The Citizen-Soldier and U.S. Military Policy, Our Vanishing Freedom: The Right to Keep and Bear Arms*, and *The Rise and Decline of the American Militia System*. He currently resides in Everett with his wife of 48 years, the former Sheila Elaine Bailey.

When Japan attacked Pearl Harbor and other Western positions in the Asia-Pacific World in December 1941, it was unprepared to go to war with the United States, and the Western Democracies generally, and even realized it could not win. Its navy and air force were impressive, and its army could battle impressively against China, but Japanese small arms were terrible. Japan's tanks could not compete with their opposite numbers. The Empire's logistical base was undeveloped for modern warfare. While the Allies could produce large numbers of trained pilots, Japan produced very few. When its elite airmen were lost at the Battle of Midway in June 1942, Japan could not replace them. At sea, Japan built battleships when it needed more aircraft carriers. The Japanese military never even attempted to win World War II by a simple and direct plan. Its planners consistently assumed that the enemy would do precisely what they assumed and countenanced no alternative analyses of facts.

Contents

Introduction

The strategy of the Imperial Japanese Navy at the outset of World War II was underpinned by several key assumptions. The most important and overriding consideration was that the war against the United States would be decided by a single decisive naval battle or Kantai Kessen. Just as the Russo-Japanese War was decided by the naval Battle of Tsushima fought on May 27–28, 1905, this great naval clash with the United States would be determined by the big guns aboard battleships. In the military strategy known as War Plan Orange, many American naval strategists in fact agreed that this one great battle would eventually take place. Japan dedicated all its naval vessels to the support of the battleships which would meet the Americans in this final battle. The Japanese correctly assumed that, at the start of any conflict, their navy would quickly seize the largely unprotected American-held Philippines, thus forcing the United States to drive across the Pacific to recapture them. The great decisive clash would inevitably occur somewhere in the western Pacific which the Imperial Japanese Navy had identified as the right area to stop the American advance.

Scenes of Japanese airplanes zeroing in on American ships in Pearl Harbor on 7 December 1941 are well known to history buffs and television watchers. As American battleships explode with huge fireballs reaching toward the heavens, the familiar voice of President Franklin Delano Roosevelt comes across denouncing "a day that will live in infamy. . . ." A narrator then reminds us that the architect of this mass destruction, Admiral Yamamoto had predicted that he could assure his divine emperor of a string of victories for six months after which there would be trouble. Therefore, the sole chance Japan had to win the Pacific War was a massive all-out sea victory that would destroy the ability of the United States to wage war in Asian waters.

Conventional wisdom among scholars of World War II asserts that it was inevitable that Japan would lose the Pacific War to the United States and the Western Allies. They largely base that conjecture on superior American military and economic power. Moreover, U.S. strategy forced the war against Japan onto a path which produced unstoppable Allied victory. Japan changed its imperialistic war on the Asian mainland into the Pacific War by simultaneously attacking Pearl Harbor, Malaya, Hong Kong, and the Philippines. Their leaders sought two outcomes by expanding the war from a regional conflict into a world war. First, they looked for access to various natural resources for Japan which had few resources of its own. Secondly, they wished to end the ongoing, and apparently never-ending, war with China that had become a proxy war with Western powers. According to conventional theory, the Japanese military expected the Allied nations to concede Japan's territorial gains in the first six months of the war, ensuring a "Greater East Asia Co-Prosperity Sphere," and thus relinquish access to the resources the Japanese desperately needed. Myers contended that there was precious little evidence supporting that conjecture.[1]

Recently, Michael W. Myers, a professor at Washington State University's School of Politics, Philosophy, and Public Affairs, published a book that counters that established historical paradigm and argued that Japan did indeed have a chance to win the war.[2] Myers took issue with conventional historians' assertion that Japanese strategy was predicated upon a strategic concept of a single decisive victory, especially a naval defeat of the United States. However, his evidence is taken more from army, not navy, commanders. For example, General Yamashita wrote to his superiors in Japan about the failure of decisive battles and the need for longer term strategy. Messages from several other Japanese military leaders also rejected the concept of the Great All-out Victory. Many Japanese army, not navy, leaders were fully aware that the Pacific War was a war of attrition. This was, however, a minority opinion in the army and a minuscule position held among naval leaders. The One Great Battle

---

[1]Barnhart, Michael A. *Japan Prepares for Total War: The Search for Economic Security, 1919-1941.* Cornell University Press, 1987.

[2]Myers, Michael W. *The Pacific War and Contingent Victory.* University of Kansas Press, 2015.

remained the wholly dominant idea of the major military decision makers.

Our overall thesis in this tome is that the Japanese deserved to lose the Pacific War because of their own conduct and behavior. We name three basic areas in which Japan was a failure. First, the Japanese sought to win the great all-out final victory at sea, but failed in the four times that such a win was possible. Second, the Japanese had modernized, but not nearly enough. Their weapons were either obsolescent, if not obsolete, at the beginning of the Pacific War, or became so as the war progressed and other nations advanced their weapons technology. Third, Japan had so utterly destroyed any vestige of humanity and morality in their armed forces that, on a supernatural level, Japan was doomed. World War II was determined not just by economic and military power, but by battles, technology, production, leadership, and morale.[3]

Our study begins with a long look at Alfred Thayer Mahan, the brilliant American geopolitical strategist from whose work, *The Influence of Sea Power upon History,* the Japanese drew their strategy. The way to establish supremacy in any theater of war is to win absolute control over the seas. All modern military and naval intelligence started with Mahan. The Japanese rendered his ideas inflexible.

In four decisive battles, the Japanese might have achieved victory in a large-scale confrontation, but they failed because of their own cowardice and lack of nerve. In the first, the sneak attack on American warships in Pearl Harbor that occurred without first declaring war, the Japanese seemed more interested in returning the fleet intact that in destroying the enemy. They left behind huge fuel supplies, dry dock and repair facilities, and the American aircraft carriers. All of the ships, save the obsolete Utah, were eventually salvaged and used against Japan. The loss of American lives was heartrending, but not damning. And the few carriers proved their worth repeatedly until American industrial might produced multiple new and improved carriers.

At Midway, the Japanese violated one of Mahan's cardinal principles and divided their forces. By making the plan so complex, they moved to insure its defeat. When the four prime carriers were destroyed, they failed both to bring in the remaining carriers and to invade Midway

---

[3]Overy, Richard. *Why the Allies Won.* Norton, 1996.

island. Indeed, the complex plan never made it clear which was the first duty: destroy the enemy carriers or secure Midway. The carriers that the Japanese did not re-deploy from the decoy strike on Alaska would have been much more likely to win against what the Japanese supposed was one remaining U. S. carrier than later when American carrier strength had grown.

At the Battle of the Philippine Sea, the Japanese chance of victory was considerably diminished. American sea power had grown while Japanese sea-power was far weaker than it was at the beginning of the Pacific War. Japan had not trained a new generation of skilled pilots and their losses were horrific. Americans referred to this as the Great Marianas Turkey Shoot.

Their last great battle actually proved to be the best opportunity for Japan since its clandestine attack on Pearl Harbor. While the ships of several of the prongs of the complex Sho-Go plan were mauled, Japan truly blew it at Leyte Gulf. One part of the plan was working better than might have been expected. Admiral William Halsey took the bait which was in the form of the remaining understaffed carriers. He chased them with his whole command, failing to create a protective force to guard the Leyte landing. Only the extreme heroism of a small group of vessels known as Taffy 3 that was assisting the ground troops stood between the vulnerable landing craft and the powerful Japanese navy. But the Japanese commander lost his nerve and from the jaws of victory snatched final defeat. Misinterpreting the tiny jeep carriers and destroyers as larger vessels, he pulled away just as this defense had collapsed. Never again would the Japanese Navy be able to challenge the U. S. Navy.

The average Japanese soldier was far less well-equipped than his American counterpart. The obsolescent Arisaka bolt-action rifles dated back to 1905, were needlessly long for the small Japanese soldier, had safeties that were extremely difficult to operate, had a bewildering set of sights, and used ammunition that was not waterproof. They had no machine guns that compared with the U. S. Thompson and Reising automatic arms. One Nambu semi-automatic pistol was so poorly designed that it could discharge when extracted from the holster. There was nothing comparable to the wonderful Browning automatic rifle [BAR]. There was no anti-tank bazooka type arm. Banzai swords may have had some psychological effect but were not especially useful in modern warfare.

Japanese so-called medium tanks were in fact light tanks by American and European standards. Initially, American tanks had difficulty knocking them out because their shells went straight though the light armor. After switching to explosive ammunition, American tanks quickly destroyed their Japanese opponents.

The Japanese Zero, arguably, was the finest fighter aircraft in the world in 1941. Most American aircraft in 1941 were dated, some severely outmoded. While the United States produced a whole new generation of superior airplanes, some specifically designed to fight the Zero, the Japanese stood pat. American planes such as the F6F Hellcat, P-51 Mustang, and P-38 Lightning were far and away superior to the Zero.

The Japanese had little regard for logistics. One estimate is that, in certain theaters such as Burma, more Japanese soldiers died of starvation and diseases related to poor diets than died in fighting the enemy. Medical care was, at best, primitive. Leather belts, slings, pouches, and packs rotted in tropical climates. Clothing was rarely altered to befit fighting conditions.

The Japanese psyche was filled with contempt for any but their own race. All other nations were inferior and thus could be mistreated without any scruples. They looked down upon most other Asians for their inability to develop industrially and technologically. It was obvious that none of the other Asians were worthy opponents in battle. The samurai spirit pervaded the soul of all the armed forces. To surrender was disgraceful. To die for the divine emperor was an honor. To endure hardships and brutal discipline was a sign of great national character.

Thus, the Japanese could, without giving the matter second thought, use human beings in the most brutal and degrading manner. They enslaved various native populations, moving them at will around conquered islands. They were especially brutal to the Koreans whose homeland they had occupied in 1910. Many Chinese and Korean women were consigned to brothels, serving as comfort women. Japanese occasionally used Chinese as victims for their sport of beheading with the samurai sword. One newspaper report tells a contest between Japanese officers to split Chinese in half.

The so-called medical experiments conducted by Unit 731 were, if anything, more barbaric than Nazi atrocities. Thousands of Chinese, along with a few Russians and even Americans, were subjected to all manner of diseases and experiments, including vivisection. The toxins developed there, along with poisons, were used to obliterate Chinese civilians as well as members of the military.

Because of their strict prohibition against surrender, Japanese truly looked down upon those soldiers who surrendered. During the infamous Bataan death march, they herded American and Filipino soldiers to internment camps without giving them adequate water, food or medical attention. The use of POWs to aid the enemy war effort was strictly forbidden, yet thousands of POWs were used to mine coal, build the Burmese railroad, and construct airstrips.

As Allied forces moved ever closer to the Japanese home islands, Allied POWs were moved by transport ships under horrid conditions. Because they were unmarked some were sunk by friendly action. In other cases, the guards killed the POWs rather than have them fall into Allied hands and relate the stories of atrocities.

By 1943, Japan was essentially defeated, not so much as a result of direct military conflict, but as a result of secondary activities and events. The Japanese troops were suffering from a shortage of supplies, especially food, medicine, munitions, and armaments, largely due to submarine interdiction of supplies, and losses to Japanese shipping. all of this was exacerbated by increasing difficulties brought about by the longstanding rivalry between the Imperial Japanese Army and the Imperial Japanese Navy. Malnutrition, even starvation, along with various tropical diseases all took enormous tolls of Japanese fighting men. As many as two-thirds of Japan's total military deaths were caused from illness or starvation rather than enemy gunfire.[4]

[4]Gilmore, Allison B. *You Can't Fight Tanks with Bayonets: Psychological Warfare against the Japanese Army in the South West Pacific.* Lincoln, Nebraska: University of Nebraska Press, 1998, p. 150.

Psychological Failure

The term "Victory Disease" has come into frequent use regarding the military history of Japan at the outset of the Pacific War. It has been defined as a complacency or arrogance, brought on by the astounding and seemingly unstoppable series of victories won by both the Japanese army and navy in the first six months of the war. The term in Japanese is "senshoubyou." It was rarely used before World War II, but since has been retroactively applied to many earlier overconfident aggressors.

Generally, victory disease can be shown to afflict military commands and armies which, after victories, become weak and susceptible to defeat. telltale signs of infection by the victor disease include: increasing arrogance of the military and political leaders; lethargy and complacency, expecting continued victories without diligent planning; continued use of previously established and proven victorious patterns of fighting rather than developing new tactics to anticipate enemy advances; stereotyping of enemies leading to the underestimating of those enemies; and, finally, ignorance of contrary intelligence or refusal to recognize it.

The natural Japanese disdain for western enemies appeared fully justified when thousands of Allied infantry surrendered all over southeast Asia. After the sneak attack on Pearl Harbor on 7 December 1941, Japan won a series of nearly uninterrupted victories against the Allies in Southeast Asia and the Pacific. Easy victories made the Japanese believe in their own invincibility, eventually leading the imperial troops to disaster. As the Allies geared up for all-out war they began to employ strategies that defeated the overconfident Japanese. Conduct of the war which had proven effective in earlier combat turned catastrophic against the now wiser enemy. The entire Japanese high command had become afflicted by "victory disease" and thus failed to anticipate that a new Allied tactics, higher troop morale, and improved equipment, produced vastly different results. The overconfident high command of both army and navy even chose to disregard such military intelligence as would have enabled them to realize that new tactics were sorely needed. Although the "victory disease" did not absolutely foretell defeat, it did precede it.[5]

Although the Japanese had originally planned only to establish a strong defensive perimeter to protect the home islands and to guarantee control over their conquests, the uninterrupted string of victories encouraged them to continue expanding, severely strained logistics and the navy.[6] The "victory disease" was rudely interrupted by the Battle of Midway in 1942, which was a catastrophic defeat of the Imperial Japanese Navy from which it never recovered.[7]

The entire Japanese nation came under the spell of its state-sponsored philosophy which held absolutely that the Japanese warrior is far and away superior to his opponent and thus more important than his weapons. Bushido originally was a code designed to protect the privileges of the elite samurai class over the oppressed general populace. It obviously supported an anachronistic notion of class privilege.[8] *The Imperial Rescript for Seamen and Soldiers,* written in 1890, defined Japan as the most "sacred nation protected by the gods."[9]

Bushidō (武士道) literally means 'the way of the warrior.' It was essentially the Samurai code, roughly analogous to the medieval European concept of chivalry. Like the European Christian, the Japanese Shinto samurai had a code to live up to. It is quite striking the many similarities between these codes. The term comes from bushi (warrior) class of medieval Japan. The Meiji Restoration began a major attempt to modernize still medieval Japan (1868). And while one of those efforts was to dismantle the samurai class, the principles of Bushido came to be the

---

[5]https://en.wikipedia.org/wiki/Victory_disease. See also https://psychology.wikia.org/wiki/Victory_disease

[6]"Land Disposal Plan in the Greater East Asia Co-Prosperity Sphere," drafted by the Research Section of the Japanese Army Ministry in December 1941. The text appears as the appendices of Richard Storry, *The Double Patriots: A Study of Japanese Nationalism.* Boston: Houghton Mifflin, 1957.

[7]https://people.com/archive/japans-mistake-vol-36-no-22/

[8]Cleary, Thomas, ed., trans. *Code of the Samurai: A Modern Translation of the Bushido Shoshinshu of Taira Shigesuke.* Tuttle, 1999.

[9]Quoted in https://en.wikipedia.org/wiki/Japanese_nationalism

fundamental basis of ethical training for the modernizing Japanese society, The Meiji Emperor replaced the feudal *daimyo* (lord) as the focus of loyalty and sacrifice for the Japanese people. Bushido contributed to the rise of extreme Japanese nationalism following the Meiji Restoration. The Bushido Code became the core of the martial spirit exhibited by the Japanese military during World War II. The Bushido spirit was a code for Samurai. The militarists, with the advent of the War, demanded that the civilian population also adopt the Bushido Code. Military officers assigned to schools preached it to the children.[10]

The *Bushido* code as a motivating force first entered the world's consciousness about 1900, with the publication of Nitobe Inazo's[11] *Bushido: The Soul of Japan.*[12] This book presented Bushido as an admirable code of chivalry, with a focus on a dedication to duty and thus inspired people around the world. Nitobe explained that Japanese children were educated on morality without religion playing a part in the Japanese educational system. In Japan, *Bushido*, not religion, imparts moral precepts upon its children. Nitobe strongly contended that one cannot possibly begin to understand the moral concepts of Japan unless one first understands Bushido and the cultural system from which it arose which was feudalism. His intention was to "revere the past and to admire the deeds of the samurai." The initial publication of this book was in English, not Japanese, because the book was intended to educate non-Japanese

[10]"Bushido" in Hunter, J. *Concise Dictionary of Modern Japanese History*. University of California Press, 1984.

[11]Nitobe Inazo 1862 –1933) was a Japanese agricultural economist, author, educator, diplomat, politician and a prolific writer. He published many scholarly books as well as books for general readers and contributed hundreds of articles to popular magazines and newspapers. Nitobe is most famous in the west for his work *Bushido: The Soul of Japan*, which was one of the first major works on samurai ethics and Japanese culture written originally in English for Western readers. The book was criticized because it portrayed the samurai in terms so Western that it took away some of the meaning.

[12]Nitobe, Inazo. *Bushido: The Soul of Japan.* 1899. Nitobe was a Buddhist who converted to Christianity. Nitobe's text is one of the first major works on samurai ethics and Japanese culture. It is considered by some to be the first collective statement of what is commonly referred to as the Seven Virtues of Bushido. This version of the expose of Bushido is considered significant because it was prepared during Japan's emergence as a modern nation.

about Japanese values. It has been read by many prominent Americans including U.S. Presidents Theodore Roosevelt and John F. Kennedy.[13] Bushido was also said to have inspired Robert Baden-Powell's Scout Movement.

In the Pacific War, Japanese officers adhered to *Bushido*, the samurai code, and did not abide by Western laws.[14] *Bushido* is linked with many facets of Japan's disastrous wartime experience. The seeds of this relationship were sown decades before the war, when Japanese textbooks were revised in 1910 and a generation was indoctrinated with nationalist interpretations of the samurai code. The Bushido ideology played a serious role in World War II in that it seemed to fuel Japan's determination to fight to the very last man and woman. This ferocious doggedness in turn influence the Allies' decision to drop nuclear bombs on Hiroshima and Nagasaki. By the 1940s, Bushido had become a byword for Japan's suicidal tendencies and military cruelty.

The myth of the warrior's code that was thrust upon the Japanese people made it extremely hard for wise heads to prevail in the face of fanatical and unrealistic optimism. Commanders were afraid to appear less enthusiastic than subordinates so the common denominator in every decision was tainted with fanatical optimism in divine intervention in favor of Japan. When the promised divine intervention failed to materialize, Japan suffered for it.

The first and most important conclusion of this Japanese mindset was that their leaders always underestimated their enemy's capabilities while grossly overestimating their own abilities. Their war planners failed to give any thought to the value of their enemy's quality, quantity, bravery, or strategic grasp. Some students of the Japanese psyche have concluded that this inability to properly estimate their enemy was no doubt developed subconsciously because Japanese leaders had concluded that the home land could not defeat strong enemies. One of the most serious short-term strategic failures in under-estimating the enemy concerned the Allies' ability to recuperate after the first Japanese blows. Japanese leadership was

---

[13]Yamamoto, Tsunetomo. *Hagakure: The Secret Wisdom of the Samurai.* Trans. Bennett, Alexander. Tuttle, 2014.

[14]Yoshikawa. *The Bushido Code*. Pocket Books, 1989.

certain that, following a decisive strike at Pearl Harbor, the United States would fold, retreat, negotiate, and accept defeat. The Japanese mindset could never begin to understand that the dastardly and cowardly sneak attack would cause the nation to unite as never before and strike back with untold vengeance.

Concurrently, the Japanese woefully underestimated the length of time it would take the Allies to launch counteroffensives. Indeed, many Japanese planners had assumed that the United States, in accepting the sneak strike as a victory, would fold so completely that it might never rebuild a navy that could defeat and then wholly dominate Japan. Perhaps, in another future generation, the U.S. planners might devote some effort to creating a Pacific fleet, but not in the current generation.

The most serious long-term strategic failure was Japan's complete inability to understand its own industrial weakness and the overwhelming industrial power of its enemies. Having created a slave labor camp in the former Chinese province of Manchuria, and having an enormous supply of Chinese and Korean laborers, Japanese leaders needed only the raw materials and energy they began to acquire throughout Southeast Asia. They anticipated no real opposition to their occupation of former European colonies.

Neither did Japan expect that the Allies could interrupt the shipments of oil and raw materials from the outlying areas of the newly conquered Japanese Empire. Not comprehending the German effort to destroy Allied shipments by submarine warfare, Japan had not developed a protective system such as the American convoys in the Atlantic. Neither had Japan coordinated air defense of its shipping routes with long-range aircraft from occupied islands.

In part because Japan did not understand the war with the United States, it could not accurately predict the ability of the U.S. to mobilize its resources. Japanese planners did not understand the breadth of the U.S. educational infrastructure and thus did not believe that the U.S. could possibly train the officers, non-commissioned officers, and specialists, let alone the vast number of combat troops, needed to take back Japan's conquests. Neither did they understand how quickly the United States could equip its troops with the best modern weapons used by any nation.

Japan's training techniques, such as naval aviator training, did not allow for the efficient transformation of the average Japanese aviator into a competent pilot. If Japan could not do it, then the racially inferior Americans could not possibly do it either. The Japanese had not adequately studied the American or British armies and therefore could not fathom their latent engineering and manufacturing capabilities.

Faulty and shoddy estimates of enemy strength originated with Japanese leaders as a result of the invasion and fighting in China. At best, there were, in 1937, a total of 10 reasonably well-trained Chinese divisions. The remaining 200 or so Chinese divisions, many really hordes gathered at the same place, had very unsatisfactory training. Some indeed, warlord troops especially, had no training at all. Chinese equipment was obsolete and poorly maintained, while ammunition was rare, with troops often having only a few rounds per man. Many units had one small arm for each two or more conscripts. Having learned their warrior trade and exercised their logistics by fighting the Chinese, the Japanese army grossly overestimated its worth. One American general depicted the Chinese Army as at best resembling medieval great militia known as the Fyrd.[15]

The Japanese Army, spoiled by easy victories against the Chinese, had little need, and thus little use, for military intelligence about the Chinese. The ever-victorious Japanese Army continued to ignore its own counter-intelligence practices, as superfluous. There seemed no need for an elaborate operational intelligence system because the Japanese easily obtained intelligence from the Chinese themselves. Japanese officers developed an utter disdain for the hapless Chinese. It was often worse than disdain; it became a virulent, decades-long, government-driven indoctrination of contempt. They viewed the Chinese as little better than animals and insects. China could easily be subdued, proving once again that the Japanese race and psyche were all-conquering. In 1937, in fact, War Minister Hajime Sugiyama had assured the Emperor that Japan could crush China within a month.

---

[15]The fyrd were working men who were called up to fight for Anglo-Saxon kings in times of danger. The leaders of the fyrd, the thegns, had sword and spears but the rest of the men were inexperienced fighters and carried weapons such as iron clubs, slings, axes, scythes, sickles and haymaking forks.

The easy victories against Allied armies and navies buttressed belief in the Bushido code and invincibility of the Japanese race. Commonly called the Victory Disease, this bolstering of the spirit emanated from the quick defeats of the Dutch in Indonesia, the British in Malaya, and the United States at Wake Island and in the Philippines. The failures of Japanese intelligence to properly estimate the size of its opponents is legendary. Japanese army officers estimated the strength of General Douglas MacArthur's army in the Philippines at 25,000 men when there were actually 80,000 combat troops. Japanese intelligence estimated British forces in Singapore at 30,000 when there were 85,000. It underestimated the Midway garrison at 750 men and 60 aircraft when there were 3,027 men and 121 combat aircraft. Yet, because of the motivation precipitated by the Bushido code and the Japanese racial superiority its army overcame the vast differences in the size of opposing forces.

The Japanese Navy fared as well. Nothing bolstered Japanese claims of armed superiority than its naval successes early in the Pacific War. The IJN had sunk two of Great Britain's most modern and best equipped ships at the Naval Battle of Malaya on 10 December 1941. The sinking of the battleship *Prince of Wales* and the battle-cruiser *H. M. S. Repulse,* without significant loss on the Japanese side, effectively ended British threat in the Indian and Pacific Oceans.

The Japanese invasion of the Dutch East Indies progressed at a far more rapid rate than IJN high command had ever anticipated. Naval forces from the Japanese-held Palau Islands and successfully occupied bases in Sarawak and the southern Philippines and also seized bases in eastern Borneo and in northern Celebes. Imperial troops on IJN transports advanced southward through the Makassar Strait and into the Molucca Sea. To oppose these invading forces the Allies assembled a small force, consisting of Dutch, American, British and Australian warship, many of which dated to World War I. On 23 January 1942, a force of four American destroyers attacked a Japanese invasion convoy in Makassar Strait as it approached Balikpapan in Borneo, achieving no clear victory. On 13 February, the Allies again fought unsuccessfully in the Battle of Palembang which was an attempt to prevent the Japanese from capturing the major oil port in eastern Sumatra. On the night of 19 and 20 February, an Allied naval force attacked a Japanese invasion force off Bali in the Battle of Badung Strait, again with inconclusive results. Japanese carrier aircraft successfully destroyed much of Darwin harbor in Australia on 19 February. IJN inflicted a most disastrous defeat on the Allied Navy, on 27 February 1942, followed by secondary actions over the next several days. Dutch Rear-Admiral Karel Doorman commanded American-British-Dutch-Australian Command (ABDACOM) Strike Force. Doorman's Combined Striking Force attempted to attack the IJN invasion convoy, but the Japanese withstood each assault, inflicting substantial losses on the Allies. Doorman was killed and most of the Allied warships were sunk or damaged. The end result was the Japanese occupation of the entire Netherlands East Indies with its vast supplies of oil and natural resources.[16]

At the first Battle of Savo Island, during the Guadalcanal Campaign, Japanese Admiral Mikawa thoroughly defeated an Allied fleet consisting of eight cruisers and fifteen destroyers under British Rear Admiral Victor Crutchley. In a night action, Admiral Mikawa thoroughly surprised and routed the Allied force, sinking one Australian and three American cruisers, while suffering only light damage in return. The battle has often been cited as the worst defeat in a fair fight in the history of the United States Navy. One must note, however, the failure of Admiral Mikawa to capitalize on his success at Savo. Having sunk four Allied cruisers. Mikawa conferred with his staff about whether they should turn to continue the battle with the surviving Allied warships and try to sink the Allied transports in the two anchorages. He was almost certainly unaware that the U.S. carriers had withdrawn from the battle area and thus could not be a threat the next day. Several of Mikawa's staff urged an attack on the Allied transports, but Mikawa ordered his ships to retire, again losing a great opportunity to strike a decisive blow and come very close to winning a great decisive naval battle.[17]

It was easy to conclude, following such an unchallenged and

---

[16]Thomas, David A. *The Battle of the Java Sea*. New York: Stein & Day, 1968.

[17]Hornfischer, James D. *Neptune's Inferno: The U.S. Navy at Guadalcanal*. New York: Bantam Books, 2011; and Kilpatrick, C. W. *Naval Night Battles of the Solomons*. Exposition Press, 1987.

uninterrupted string of victories, that the gods favored Japan. What patriotic Japanese would dare to question that divine inspiration?[18]

Education

Prior to the beginning of the Pacific War, the state had begun to thoroughly indoctrinate school children with a militaristic education. Teachers treated war as a science with a natural life and moral justification of its own. Because Japan was a pure nation, its wartime objectives and methodologies could not be questioned. Authorities expected its educators to teach that the Japanese were a racially, as well as morally, superior race. Historians must show overwhelming pride in Japan. Japan was the sole distinguished nation, existing as a distinct entity above all others.

Military nationalism in Japan was known as *bushidō,* that is, the way of the warrior. The term is used to describe a coherent code of beliefs and doctrines about the proper path of the samurai, or what is called generically warrior thought. Credit may properly be given to Yamaga Sokō (1622-1685) for developing and popularizing the Japanese military ethos.[19]

Japanese nationalism taught that the Japanese are a monolithic nation with a single immutable culture which promoted the cultural unity of the Japanese. It encompassed a broad range of ideas and sentiments obtained in whole or in part from many native and foreign writings. Emphasis was placed most strongly upon compatible ideas and myths which the Japanese nation had harbored over the previous two centuries regarding their native country, its cultural nature, political form and historical destiny. This new political nationalism went far beyond traditional Japanese cultural nationalism. Indeed, many varieties of cultural nationalism, such as those associated with folkloric studies, were necessarily hostile to emerging state-fostered nationalism. A relatively

---

[18]Newcomb, Richard F. 1961. *The Battle of Savo Island.* New York: Henry Holt, 1961. This is a brutally honest history of the disaster that was the battle of Savo Island, attributing most fault to Crutchley.

[19]Yamaga Sokō was a Japanese philosopher and strategist under the Tokugawa shogunate. As a scholar he applied the Confucian idea of thc “superior man” to the samurai class of Japan. This became an important part of the samurai way of life and code of conduct.

new and state-created ideology it became a politically useful nationalism, known as Shōwa statism.[20]

About the same time as Japan attacked Pearl Harbor, the state renamed elementary schools as National Schools and charged their teachers to produce children dedicated to the Emperor, future warriors who would be prepared to sacrifice themselves for the nation. Children learned to march on their way to school. A significant portion of school time was dedicated to nationalist indoctrination; to instruction on loyalty and dedication to the emperor; and to inculcating virtues such as frugality, self-sacrifice, honesty, and diligence. The state required teachers to instruct using Japanese science which was based on the Imperial Way of life. Japanese science precluded teaching evolution which challenged Japanese claims to have divine descent. Students were required to perform community service as a way to learn dedication to the Emperor. Schools based various compositions, drawings, calligraphy, and pageants on military themes. Those who left school after completing six years were still required to attend night school to learn ever more about glorious Japanese history and ethics. There was also continued military training for boys, and home economics for girls. As the war progressed, the state required teachers to place ever more emphasis on the children's destiny as warriors.[21]

Much of wartime Japanese education was based upon the writings of Shūmei Ōkawa (1886-1957), a right-wing political philosopher. In 1926 Ōkawa published *Japan and the Way of the Japanese*, which helped popularize the concept of the inevitability of a clash of civilizations between Japan and the west. Politically, his theories emphasized that Japan needed to return to its traditional kokutai traditions[22] in order to survive the increasing social tensions created by industrialization and foreign cultural influences. Following Ōkawa, the state made great efforts to

---

[20]Benedict, Ruth. *The Chrysanthemum and the Sword: Patterns of Japanese Culture*. Mariner Books, 1954.

[21]Bix, Herbert. *Hirohito and the Making of Modern Japan,* Harper Perennial, 2001.

[22]Kokutai is a concept in the Japanese language which is difficult to translate, but may be viewed as "system of government," "sovereignty," "national identity, essence and character," "national polity; body politic; national entity; basis for the Emperor's sovereignty; Japanese constitution."

foment a Japanese spirit, especially in popular culture.[23]

In 1937, in preparation of all-out war, the government issued *Cardinal Principles of the National Body and Structure,* a pamphlet of 156 pages. This publication was an official publication of the Japanese Ministry of Education, circulated with millions of copies disseminated throughout the home islands and the empire. It contained the official teaching of the Japanese state on every aspect of domestic policy, international affairs, culture, and civilization. The most important purpose was to overcome social unrest and to develop a new Japan spirit of self-sacrifice.[24] School pupils were taught to put the nation before the self, and that they were part of the state and not separate from it.[25] There is no existence for an individual outside the state; all exists within and for the polity It also instructed them in the principle of *hakkō ichiu,* which would be used to justify imperialism.[26]

Baron Sadao Araki (1877-1966) espoused the doctrine of *seishin kyōiku,* that is, spiritual training, for the ideological indoctrination of military personnel. In this way he led the politicization of the Japanese Army. As Minister of Education, he fostered the integration of the samurai code into the national education system.[27]

---

[23] Wakabayashi, Bob. *Modern Japanese Thought.* Cornell University Press, 2007.

[24] Jansen, Marius B. *The Making of Modern Japan.* Belknap Press, 2000; Samuels, Richard J. *Securing Japan: Tokyo's Grand Strategy and the Future of East Asia.* Cornell University Press, 2007.

[25] McClain, James L. *Japan: A Modern History.* New York: Norton, 2001, pp 465-6, 470. See also Beasley, W. G. *The Rise of Modern Japan.* Palgrave Macmillan, 1990, p 187.

[26] Hakkō ichiu may be roughly translated as "eight crown cords, one roof" i.e. "all the world under one roof." It was a Japanese political slogan that became popular from the Second Sino-Japanese War to World War II, and was popularized in a speech by Prime Minister Konoe on January 8, 1940.

[27] Baron Sadao Araki was a general in the Imperial Japanese Army before and during World War II. He was one of the principal nationalist right-wing political theorists in the Empire of Japan, and was regarded as the leader of the radical faction within the politicized Imperial Japanese Army and served as Minister of War under Prime Minister Inukai. He later served as Minister of Education during the Konoe and Hiranuma administrations. At the Tokyo war crimes trials following World War II, he was convicted of war crimes and given a life sentence.

Alfred Mahan and Geopolitics

Alfred Thayer Mahan was a United States Navy flag officer, geostrategist, geopolitical theorist, and historian, who has been called "the most important American strategist of the nineteenth century." His concept of "sea power" was based on the idea that countries with greater naval power will have greater worldwide impact; it was most famously presented in his 1890 book *The Influence of Sea Power Upon History, 1660–1783.* The concept had an enormous influence in shaping the strategic thought of navies across the world, especially in the United States, Germany, Japan and Britain, ultimately causing the World War I naval arms race. His ideas still permeate the U.S. Navy Doctrine as well as that of many other nations.

Alfred Thayer Mahan was born on 27 September 1840, West Point, New York, a son of Dennis Hart Mahon (1802-1871) and his wife Mary Helena Okill (1815-1893). He married Ellen Lyle (1851-1927). Dennis Mahan was a professor at the U. S. Military Academy at West Point. Alfred graduated from the U.S. Naval Academy at Annapolis, Maryland, in 1859. He served almost 40 years in the U. S. Navy. He fought in the Civil War, later served on the staff of Admiral J. A. B. Dahlgren, and progressed steadily in rank. In 1884, he was invited by Stephen Luce, president of the newly established Naval War College at Newport, Rhode Island, to lecture on naval history and tactics. Mahan served as president of the Naval War College from 1886 to 1889 and again from 1892 to 1893. He retired as rear admiral in 1896, but returned to the navy during the Spanish-American War to serve on the Naval War board. Mahan retired from the U.S. Navy in 1896 but was subsequently recalled to service. In T*he Interest of America in Sea Power, Present and Future* (1897), he sought to arouse his fellow Americans to a realization of their maritime responsibilities. Mahan served as president of the American Historical Association in 1902. His other major books include *The Life of Nelson*

(1897) and *The Major Operations of the Navies in the War of American Independence* (1913). Before his death in December 1914, Mahan correctly foretold the defeat of the Central Powers and the German navy in World War I. He died on 1 December1914, in Quogue, New York, age 74, and was buried at Quogue Cemetery.[28]

Admiral Mahan was an American naval officer and historian who was a highly influential exponent of sea power in the late 19th and early 20th centuries. In 1890 Mahan published his college lectures in book form entitled *The Influence of Sea Power upon History, 1660–1783*. In his book he made out a case for the paramount importance of sea power in order to establish national historical supremacy. The appearance of the book was most timely because it was published at a time when there was great technological improvement in warships. It won immediate recognition abroad, leading to a second book, *The Influence of Sea Power upon the French Revolution and Empire, 1793–1812* (1892). Both books were avidly read in Great Britain and Germany, where they greatly influenced the buildup of naval forces in the years prior to World War I.

In his memoirs, *From Sail to Steam*, Mahan credited his reading of Theodore Mommsen's six-volume History of Rome for the insight that sea power was the key to global predominance. In *The Influence of Sea Power Upon History,* Mahan reviewed the role of sea power in the emergence and growth of the British Empire. In the book's first chapter, he described the sea as a "great highway" and "wide common" with "well-worn trade routes" over which men pass in all directions. He identified several narrow passages or strategic "choke-points," the control of which contributed to Great Britain's command of the seas. He famously listed six fundamental elements of sea power: geographical position, physical conformation, extent of territory, size of population, character of the people, and character of government. Based largely on those factors, Mahan envisioned the United States as the geopolitical successor to the British Empire.

In the first book, Admiral Mahan asserted that British control of

---

[28]Find A Grave Memorial 7972481. See also Seager II, Robert, ed. *Letters and Papers of Alfred Thayer Mahan* (3 vol 1975) v. 1. 1847–1889. – v. 2. 1890–1901. – v. 3. 1902–1914.

the seas, combined with a corresponding decline in the naval strength of its major European rivals, paved the way for Great Britain's emergence as the world's dominant military, political, and economic power. Mahan had shown the interdependence of the military and commercial control of the sea and asserted that the control of seaborne commerce can determine the outcome of wars.

Admiral Mahan believed that national greatness was inextricably associated with the sea, with its commercial use in peace and its control in war; and he used history as a stock of examples to exemplify his theories, arguing that the education of naval officers should be based on a rigorous study of history. He emphasized the locating and obtaining strategic locations. If these could not be allied in some way, then the navy must decide how to isolate or capture these points should war come.

The primary mission of a navy is always to secure the command of the sea, thus ensuring the maintenance of sea communications for one's own ships while denying their use to the enemy. Nations must also closely supervise neutral trade. Mahan believed that good political and naval leadership was important but a nation must pay great attention to geography when it came to the development of sea power. His recognition of the influence of geography on strategy was tempered by a strong appreciation of the power of contingency to affect outcomes.

The concentrated study of history and geography would aid a nation in determining which overseas properties it should obtain. The nation should move aggressively in the acquisition of important properties during peacetime. States must also increase production and shipping capacities while at peace. Finally, states must constantly upgrade and improve their navies before wars occur both to retain great strength and to be able to afford the very finest without cost related compromises.

Eight years before the Spanish-American War resulted in the United States becoming a world power with overseas possessions, Mahan wrote an article in the *Atlantic Monthly* entitled "The United States Looking Outward," (1890). In the article, he urged U.S. political leaders to recognize that our own security and interests were affected by the balance of power in the Eurasian land mass. Mahan reasoned that the United States, like Great Britain, was geopolitically an island lying offshore the Eurasian landmass whose security could be threatened by a hostile power or alliance of powers that gained effective political control of the key power centers of Eurasia. He further understood that predominant Anglo-American sea power in its broadest sense was the key to ensuring the geopolitical pluralism of Eurasia. He famously wrote in *The Influence of Sea Power upon the French Revolution and Empire* that it was only the navy of Great Britain, which he labeled "those far distant storm-beaten ships," that stood between Napoleon and the dominion of the world.

Mahan's book influenced some of the leading American politicians who came to believe that the lessons he taught could be applied to U. S. foreign policy, particularly in the quest to expand U.S. markets overseas. The publication of Mahan's books preceded much of the disorder associated with the economic downturn of the 1890s, but his work resonated with many leading intellectuals and politicians who were concerned by the political and economic challenges of the period and the declining lack of economic opportunity on the American continent.[29]

Mahan's books complemented the work of one of his contemporaries, Professor Frederick Jackson Turner, who is best known for his seminal essay of 1893, *The Significance of the Frontier in American History*. As a history professor at the University of Wisconsin, Turner had argued that westward migration across the North American continent and the country's population growth had finally closed the American frontier. The end of the frontier had profound social and economic consequences. While Turner did not explicitly argue for a shift towards commercial expansion overseas, he did realize that the nation's vigorous foreign policy showed that Americans were looking outside the continental United States for new markets. Mahan was one of many American leaders who infinitely preferred free trade to attempting autarky which they considered impossible and leading only to folly.[30]

Mahan was second to none as an exponent of a vigorous foreign

---

[29]Cowl, Philip A. "Alfred Thayer Mahan: The Naval Historian," in Paret, Peter, Gordon A. Craig, and Felix Gilbert, eds. *Makers of Modern Strategy from Machiavelli to the Nuclear Age* (1986), ch. 16.

[30]Seager, Robert, *Alfred Thayer Mahan: The Man and his Letters*. Annapolis: Naval Institute Press, 1977.

policy. Mahan argued that the U.S. economy would soon be unable to absorb the massive amounts of industrial and commercial goods being produced domestically. Thus, the United States must seek new markets abroad. Would the government guarantee access to these new international markets? Securing such access would require three things: a merchant navy, which could carry American products to new markets across the great highway of the high seas; an American navy complete with the most modern battleships to deter or destroy rival fleets; and a network of naval bases capable of providing fuel and supplies for the enlarged navy, and maintaining open lines of communications between the United States and its new markets.[31]

There are actually several interrelated theses in Mahan's book. The primary one is the assertion that naval power is a deciding factor for everyone but the most landlocked of countries. The secondary one is not so clearly stated, but quite evident in the later parts of the book, that the proper goal of military operations is the reduction of organized enemy forces in the field. The later parts of the book particularly talk about this, showing that the French government and navy held to a theory that pursued the taking of objectives while preserving force, and that it time and again failed to gain results, while the British habit of forcing battles inevitably put their opponents into a worse position over time. Lastly, he considers the pursuit of interrupting merchant shipping to be a mistaken strategy, as British trade increased even during wars where the French captured large numbers of British merchants. Mahan reviewed the successive moves toward European continental hegemony by the Spanish and Austrian Hapsburgs, Louis XIV's France, and Revolutionary and Napoleonic France, and the great coalitions, supported by sea power, that successfully thwarted those would-be hegemons.

Mahan's book covered the Age of Sailing Ships, concerned primarily with history from the Restoration of English King Charles II to the end of the Revolutionary War. However, there also was an extended chapter that considered naval power throughout history. But it is the age of sailing ships that is definitely his preferred era for him. He is not static

[31]Hattendorf, John B., ed. *The Influence of History on Mahan.* Naval War College Press, 1991.

in his thinking because he realized that tactics must change over time, with new technology. However, it is nonetheless possible to find strategic truths that always apply. His narrative gets steadily more detailed as it progresses, with the last few chapters \analyzing the actions in India and the Caribbean from 1781-1783 in considerable detail. As his descriptions get more detailed, so too do the conclusions that he draws from them.

Mahan was the first, and remains the foremost, geopolitical thinker that the United States has produced. His writings are wholly based on a profound geopolitical insight based on an understanding of the impact of geography on history. Much of the military action of the 20th century can be traced to a handful of geopolitical writers, notably Halford John Mackinder and Karl Haushofer.[32]

As early as 1904, Halford John Mackinder (1861-1947), had argued simply that the land of the planet Earth was divisible into: first, the World-Island, comprising the interlinked continents of Europe, Asia, and Africa. This was the largest, most populous, and richest of all possible land combinations; second, the offshore islands, notably Great Britain and Japan; and, third, the outlying islands, including the continents of North America, South America, and Australia. The Heartland of the World Island included land stretching from the Volga to the Yangtze and from the Himalayas to the Arctic. Mackinder's Heartland was the area then ruled by the Russian Empire and after that by the Soviet Union. In 1919, Mackinder summarized his theory as:[33]

> Who rules East Europe commands the Heartland;
> who rules the Heartland commands the World-Island;
> who rules the World-Island commands the world.

Dr. Professor General Karl Haushofer (1869-1946) was a German geopolitical thinker turned politician. Through his student Rudolf Hess, Haushofer's ideas influenced the development of Adolf Hitler's expansionist strategies. Later, Haushofer fell into disfavor with the Nazis and was confined to a concentration camp. Although Haushofer denied direct influence on the Nazi regime, and was never a member of the

[32]Sempa, Francis P. *Geopolitics: From the Cold War to the 21st Century.* Transaction Books, 2002.

[33]Mackinder, Halford. *Democratic Ideals and Reality*, p. 150.

NSDAP, he was so overwhelmed by the near annihilation of Germany that he and his wife committed suicide in 1946.[34]

German geopolitik, as expounded by Haushofer, advocates adding living space [*Lebensraum*] that would allow the growth of Greater Germany. Haushofer added older German ideas, including an organic or anthropomorphized conception of the state, and the need for complete autarky, that is, self-sufficiency through the top-down organization of society. The root of uniquely German geopolitik rests in the writings of Haushofer who justified *Lebensraum*, at the cost of other nations' existence because conquest was a biological necessity for a state's growth. Smaller and less powerful states had no right to exist and strong states could absorb their land and overwhelm their population.[35]

Haushofer had little use for politics, believing that national heritage was alone important. He advocated an alliance of Germany and Russia to control Mackinder's heartland of the world island. Haushofer, Karl. [36] He also advocated the German-Japanese alliance.[37]

Mahan's geopolitics had a direct bearing on American foreign policy. In the United States, Mahan's emphasis upon the acquisition of naval bases was as analytical as prophetic. U. S. Secretary of State William Seward had expanded the U.S. commercial presence in the Pacific Ocean by purchasing Alaska in 1867, and by exerting American influence over Hawaii. In the latter case the United States concluded a treaty that sought to bind the island's economy to that of the United States. In 1893, Mahan wrote a letter to the editor of the New York *Times* in which he recommended U.S. annexation of Hawaii as a necessary first step to

---

[34]Jacobsen, Hans-Adolf. Karl Haushofer: Leben und Werk. 2 vols. Harald Boldt Verlag, Boppard 1979.

[35]Heske, Henning, "Karl Haushofer: His Role in German Politics and in Nazi Politics," *Political Geography* 6 (1987), pp. 135–144. See also Dorpalen, Andreas. *The World of General Haushofer: Geopolitics in Action*. New York: Farrar & Rinehart, 1942.

[36]Haushofer, Karl. *Der Kontinentalblock: Mitteleuropa, Eurasien, Japan,* Berlin: Eher, 1941.

[37]Haushofer, Karl. *Das Japanische Reich in seiner geographischen Entwicklung.* Vienna: L.W. Seidel & sohn, 1921. See also Spang, Christian W., *Karl Haushofer und Japan. Die Rezeption seiner geopolitischen Theorien in der deutschen und japanischen Politik,* Munich: Iudicium, 2013.

exercise control of the North Pacific. If the United States failed to act, Mahan warned, "the vast mass of China . . . may yield to one of those impulses which have in past ages buried civilization under a wave of barbaric invasion." Should China "burst her barriers eastward," he wrote, "it would be impossible to exaggerate the momentous issues dependent upon a firm hold of the [Hawaiian] Islands by a great civilized maritime power."

Seward also attempted to purchase suitable Caribbean naval bases. Seward negotiated a treaty with the Colombian Government that would allow the United States to build an isthmian canal through the province of Panama. Congress, however, had become preoccupied with Reconstruction in the South, and the Senate ignored Seward's best efforts to create a network of American naval bases in the Atlantic Ocean. As it was, bases were not as important in the Atlantic Ocean as in the far larger Pacific Ocean.

Although President William McKinley showed little interest in Mahan's ideas, his Assistant Secretary of the Navy, Theodore Roosevelt, and Secretary of the Navy Herbert Tracy were avid supporters of Mahan's ideas. As president, Theodore Roosevelt wrote a naval history of the War of 1812[38] cast in Mahan's framework.[39]

After the outbreak of hostilities with Spain in May 1898, President William McKinley finally secured the annexation of Hawaii by means of joint resolution of Congress. Following the successful conclusion of the Spanish-American War in 1898, the United States gained control of territories that could serve as the coaling stations and naval bases that Mahan had discussed, such as Puerto Rico, Guam, and the Philippines. Five years later, the United States obtained a perpetual lease for a naval base at Guantanamo Bay, Cuba.

German naval strategist Admiral Ludwig Wilhelm Carl Borckenhagen (1850-1917) introduced Mahan's theories to the world in a

---

[38]Theodore Roosevelt, *The Naval War of 1812* (1882). Reprint by Modern Library, 1999.

[39]Turk, Richard W. *The Ambiguous Relationship: Theodore Roosevelt and Alfred Thayer Mahan.* Greenwood Press, 1987. See also Wimmel, Kenneth, *Theodore Roosevelt and the Great White Fleet: American Seapower Comes of Age.* Washington: Brassey's, 1988.

series of influential and widely read and translated papers. Thus, Mahan's name became familiar in the German navy. Always interested in military affairs, Kaiser Wilhelm II ordered his officers to read Mahan. German Grand Admiral Alfred von Tirpitz (1849–1930) used Mahan's reputation to build a powerful surface fleet. But the course of World War I changed European ideas about the place of the navy. As it was the German fleet engaged in a major, if not decisive, battle in May-June 1916 known as Battle of Jutland. It was heavy on maneuvering ships about and light on fighting. The German fleet withdrew before there was a true great all-out Mahan style battle.

The Influence of Seapower Upon History, 1660–1783 was translated into Japanese and was used as a textbook in the Imperial Japanese Navy (IJN). That usage strongly affected the IJN's plan to end Russian naval expansion in the Far East, which culminated in the Russo-Japanese War (1904–1905). While junior naval officers were studying the implications of advancing weapon technology while simultaneously harboring increasing doubt about the relevance of battleships, the senior officers remained overwhelmed by the ideas of Borkenhagen and Mahan. In the context of bureaucratic struggles between the army and navy to gain supremacy in conducting the Pacific War, the ghost of Mahan hung heavily over the Japanese naval leaders. As these men prepared for war against the United States, they made decisions that were largely based on miscalculations about American and Japanese strengths and American intentions, but usually grounded in preconceptions based on Mahan.[40]

It was the IJN's pursuit of the one great decisive battle that contributed to Imperial Japan's defeat in World War II, because the development of the submarine and the aircraft carrier, combined with advances in technology, largely rendered obsolete the doctrine of the decisive battle between fleets[41] It was British admiral Jacky Fisher who reversed Mahan by utilizing technological change to propose submarines for defense of home waters and mobile battle cruisers for protection of distant imperial interests.

Control of the sea could be achieved not by destruction of

[40]Asada, Sadao. *From Mahan to Pearl Harbor*. Naval Institute Press, 2006.
[41]Peattie, Mark and David Evans, *Kaigun*. U.S. Naval Institute Press, 1997.

commerce but only by destroying or neutralizing the enemy fleet. Such a strategy called for the concentration of naval forces composed of capital ships, not too large but numerous, well-manned with crews thoroughly trained, and operating under the principle that the best defense is an aggressive offense. His theories, expounded before the submarine became a serious factor in warfare, delayed the introduction of convoys as a defense against German U-boats during World War I. But more importantly, his ideas precluded Japanese deployment of the best protective strategies to ensure safe transport of vital goods from its conquered territories back to the home islands. By the 1930s, the U. S. Navy had deployed long-range submarines which successfully raided Japanese shipping in World War II. Again showing their inflexibility the Japanese, still tied to Mahan, designed their submarines as ancillaries to the fleet and failed to attack American supply lines in the Pacific or adequately protect merchant shipping.

The idea of the great mid-Pacific battle, a titanic clash of battleships, dominated Japanese thinking, following Mahan. It apparently did not occur to the Japanese that the United States would not be willing to engage in such a climatic, all-out battle. Japan had built its fleet with such a conflict in mind. Indeed, even after aircraft carriers had shown their importance, Japan built two super-battleships that would have spearheaded its fleet in this final battle.

Moreover, the IJN did not adhere strictly to Mahanian doctrine because its forces were often tactically divided, particularly during the attack on Pearl Harbor and the Battle of Midway. It might easily have brought all its carriers to Pearl Harbor and continued flying sorties against American installations there until both ships and land-based facilities had been utterly destroyed. It could have assigned a certain number of aircraft to seek out and destroy the enemy carriers that were at sea when the initial attack began.

Mahan argued for a universal principle of concentration of powerful ships in home waters and minimized strength in distant seas. British addendum to Mahan suggested that the home islands might be protected by submarines while powerful surface fleets roamed the high seas. Japan did not adhere to this change.

In his later articles and books, Mahan accurately envisioned the political struggles of the 20th and 21st centuries as expressions of geopolitical considerations. In *The Interest of America in International Conditions* (1910), Mahan foresaw the then-emerging First World War and the underlying geopolitical conditions leading to the Second World War, recognizing that Germany's central position in Europe. At the time Germany possessed unrivaled industrial and military might on the continent. But it was her quest for sea power that posed a threat to Great Britain. Mahan argued that this emerging desire for power on the high seas would also ultimately become a grave threat to the United States. "A German navy, supreme by the fall of Great Britain," he warned, "with a supreme German army able to spare readily a large expeditionary force for over-sea operations, is one of the possibilities of the future." "The rivalry between Germany and Great Britain today," he continued, "is the danger point, not only of European politics but of world politics as well." It remained so for 35 years.[42]

Mahan's sharp mind also grasped as early as 1901 the fundamental geopolitical factors that were to become the Cold War. These realities would soon emerge from the remnants of the first two world wars. In *The Problem of Asia*, Mahan urged statesmen to "glance at the map" of Asia and note "the vast, uninterrupted mass of the Russian Empire, stretching without a break . . . from the meridian of western Asia Minor, until to the eastward it overpasses that of Japan." He envisioned an expansionist Russia needing to be contained by an alliance of the United States, Great Britain, France, Germany, and Japan, which is precisely what happened between 1945 and 1991.

Mahan foresaw the potential power of an emerging China. He also envisioned a time when the United States would be compelled to be concerned with the power of China. In *The Problem of Asia*, Mahan depicted a future struggle for power in the area of central Asia he called the "debatable and debated ground," and identified the "immense latent force" of China as a potential geopolitical rival. "[I]t is scarcely desirable," Mahan wrote, "that so vast a proportion of mankind as the Chinese

---

[42]Mahan, Alfred Thayer. *The Interest of America in International Conditions* (1910). Reprint by Transaction Books, 2003.

constitute should be animated by but one spirit and moved as a single man." Mahan understood that Western science and technology would become globalized. Japan was merely the first Asian nation to accept the industrial and technological supremacy of the West and move to emulate them. Mahan wrote that under such circumstances "it is difficult to contemplate with equanimity such a vast mass as the four hundred millions of China concentrated into one effective political organization, equipped with modern appliances, and cooped within a territory already narrow for it."[43] He wrote that the United States must present,

> a policy realized in our past action, and to affirm it for the present and future. Our people have not now to evolve a policy, but to decide whether that of the past justifies itself to their conscience, to their sense of right and wrong, and embodies their purpose of the present. This still existent policy may, I apprehend, fairly be stated to be the determination to have equal commercial privileges, and withal to respect to the utmost the integrity of Chinese territory, and the individuality of the Chinese character in shaping its own government and polity. We meddle not with their national affairs until they become internationally unendurable. But in the very enunciation of this policy we are confronted by the fact that it is diverse from that of some other states, as shown by their acts in special instances, and plausibly to be inferred from their general course and obvious tendency. Such divergence is not always necessarily a cause for alarm, but it is for watchfulness; and it must be taken into account, as an element influential upon our own policy, not perhaps in general conception or as towards China, but in the matter of deciding upon the preparation we need, and the free-handedness to be maintained in external relations of lesser importance. Needless external preoccupations might greatly embarrass us, in case divergence from our policy should develop into opposition to our interests, or to those of civilization in general. Briefly, we cannot be sure of the commercial advantages known as the open door, unless we are prepared to do our share in holding it open.

Like Germany before the First World War, and Japan before World War II, China in the 21st century has embraced Mahan. Naval War College

---

[43]Mahan, Alfred Thayer, *The Problem of Asia and Its Effect Upon International Policies* (1905) Reprint by The Classics, 2013.

professors have examined the writings of contemporary Chinese military thinkers and strategists in this regard in their important work, *Chinese Naval Strategy in the 21st Century: The Turn to Mahan.*[44] With regard to Mahan's elements of sea power, China is situated in the heart of east-central Asia and has a lengthy seacoast, a huge population, a growing economy, growing military and naval power, and, at least for now, a stable government. China's political and military leaders have not hidden their desire to supplant the United States as the predominant power in the Asia-Pacific region. Under these circumstances, China's embrace of Mahan is reason enough for Americans to reacquaint themselves with the writings of that great American strategic thinker.

The outcome of the Pacific war was directly related to the ability of the Americans to adjust strategy to the realities of the Pacific War. The Japanese after initial success did little to adjust their strategy to the realities of the Pacific War. They refused to adapt their projected goals to allow for realistic victories. The Japanese were hamstrung by inter-service rivalries along with inadequate industrial capacity and limited natural resources.

The Japanese and the Americans each had war plans in place for the Pacific campaign. The United States had long worked on, perfected, and adapted Plan Orange since the early part of the 20th Century, that is, following the Spanish-American War and Russo-Japanese War. Predicated on holding the Philippines until relief could arrive Orange assumed that the US Pacific Fleet would sail across the Pacific and fight the Japanese Navy in a manner written about by Alfred Thayer Mahan.

When the Japanese attacked Pearl Harbor as well as the Philippines and Southeast Asia, defeating American and Allied forces in detail, crippling the American Navy and dooming, the Philippines the Americans were able to adjust strategy to first a defensive one supplemented by raids against the Japanese perimeter by carrier forces and the beginnings of a nascent submarine campaign against Japanese merchant shipping. The Americans were able to parry the Japanese thrust at the Coral Sea and inflict a major defeat on the Japanese Carrier Forces at Midway prior to launching the first limited offensive by the Navy and

---

[44]Toshi Yoshihara, Toshi and Holmes, James, *Chinese Naval Strategy in the 21st Century: The Turn to Mahan. Taylor & Francis, 2009.*

the Marines at Guadalcanal.[45]

Simply stated, the Japanese were conflicted. The Imperial Japanese Navy desired a campaign that would destroy the American Navy and expand the Empire to the East, into Southeast Asia and the Indian Ocean. The Army was obsessed with China, having been embroiled there since the early 1930s. In addition, other Japanese Army leaders had designs on Siberia and fought a brief campaign against the Soviets which had ended in a disastrous defeat. The Japanese remained mired in their conflicting strategies with the Navy primarily fighting the Pacific campaign aided by limited Army and Army Air Forces on the islands Japan had occupied or fortified while the bulk of the Army was engaged in China, Southeast Asia.[46]

---

[45] Toland, John. *Rising Sun: The Decline and Fall of the Japanese Empire 1936-1945. Random House, 1970.*

[46] Ronald Spector. *Eagle Against the Sun: The American War Against Japan.* Free Press, 1984.

The First Great Battle: Pearl Harbor

The first great all-out battle of what the Japanese called the Pacific War was the sneak attack on American naval installations on Pearl Harbor. While it is true that Japan rarely, if ever, preceded an attack on a non-belligerent power, they had gotten by largely because the nations they attacked were minor powers. Save for the odd incident with the Soviet Union, in which the Japanese were dealt a stunning blow, all the other pre-World War II unprovoked attacks had been against essentially helpless states.[47]

By 1935, in keeping with the evolution of its theoretical battle with Japan, called in the U. S. War Plan Orange, the Navy had abandoned any thought of charging across the Pacific towards the Philippines at the outset of a war with Japan. In 1937, the U. S. Navy had further determined that bringing the fleet up to wartime levels could not be accomplished in less than six months. Moreover, there was a myriad of other logistic considerations needed to execute a trans-Pacific movement. Realistic preparations simply did not exist and would require two years to execute after the onset of a war. In 1940, Chief of Naval Operations, Admiral Harold Stark had prepared a Plan Dog memorandum, which emphasized a defensive war in the Pacific while the United States concentrated on defeating Nazi Germany first. This decision consigned Admiral Husband Kimmel's Pacific Fleet to the role of keeping the Imperial Japanese Navy out of the eastern Pacific and away from the shipping lanes to Australia.

The U.S. Pacific Fleet had been stationed at Pearl Harbor since

---

[47]The latest study on the attack on Pearl Harbor offers a fascinating critique of the Japanese attack on Pearl Harbor from an operational analysis perspective. Makes a convincing case that the Japanese attack was a stunning tactical success only because the Americans were woefully unprepared and the Japanese astonishingly lucky. Zimm, Alan D. *Attack on Pearl Harbor: Strategy, Combat, Myths, Deceptions.* Philadelphia: Casemate, 2011.

April 1940. In addition to nearly 100 naval vessels, including 8 battleships, there were substantial military and air forces. U.S. Navy Admiral Husband E. Kimmel and Army Lieutenant General Walter C. Short shared command at Pearl Harbor,

By late 1941 Prime Minister Tōjō Hideki decided on war. Admiral Yamamoto Isoroku, the commander in chief of Japan's Combined Fleet, had planned the attack, acting on specific orders, against the U.S. Pacific Fleet. He had drawn upon the finest minds in the Japanese Navy to plan, critique, prepare, and execute the order. The thought was that once the U.S. fleet was out of action, the way would be open for the unopposed Japanese conquest of all of Southeast Asia along with the Indonesian archipelago. The order for the assault was issued on 5 November 1941, and on 16 November the task force began gathering near the Kuril Islands. Commanders were instructed that the fleet might be recalled, however, in case of a favorable outcome of the negotiations in Washington, D.C. Beginning on 26 November, Vice Admiral Nagumo Chuichi led his fleet including 6 aircraft carriers, 2 battleships, 3 cruisers, and 11 destroyers to a point some 275 miles north of Hawaii. From there about 360 planes in total were launched.[48]

As the tension mounted, the military in Washington issued warnings to the U. S. commanders concerning the possibility of war. Specifically, on 16 October, and again on 24 and 27 November, the high command sent urgent messages. The notice of 27 November to Kimmel began, "This dispatch is to be considered a war warning," and continued that "negotiations have ceased," and directed the admiral to "execute an appropriate defensive deployment." Kimmel also was ordered to "undertake such reconnaissance and other measures as you deem necessary." The communication of the same day to Short declared that "hostile action is possible at any moment" and, like its naval counterpart, urged "measures of reconnaissance."

Admiral Kimmel's intelligence officers had not been able to locate substantial elements in the Japanese first-line ships in the two major carrier divisions. He lacked the airplanes to substantially expand his

[48]Peattie, Mark R. and Evans, David C., *Kaigun: Strategy, Tactics, and Technology in the Imperial Japanese Navy*, Naval Institute Press, 1997.

reconnaissance activities to the northwest, the logical point for an attack. He moored the entire fleet, excepting only those ships that were at sea, in the harbor and permitted a part of his personnel to go on shore leave. Neither of these officers suspected that the base at Pearl Harbor would itself be subjected to attack. In the 10 days between the war warning of 27 November and the Japanese attack itself, no additional action was taken by Washington.

The cost of what Franklin Roosevelt called a sneaky and dastardly unproved attack would be high. Not only would the United States be outraged but so would its great Western Allies, notably Great Britain and France. If the results of this costly attack were great enough then the end might justify the means.

In terms of Mahan's thought, here was an opportunity for the great all-out battle and the Japanese could choose the time for that event that was most advantageous to them. But it must require that the virtually entire Japanese Navy be brought to bear. It could not be justified if all that happened was a hit and run one day long operation. It could be justified if the IJN destroyed much of the U. S. Navy's firepower in a single blow and followed that with the destruction of what remained, ships and facilities. Precisely, the operation could not succeed as a hit and run attack that did some damage and allowed the U.S. the wherewithal to recover and eventually strike back.[49]

On Sunday, December 7, Washington learned from intercepted and decrypted diplomatic messages that the Japanese ambassadors had been instructed to ask for an interview with the Secretary of State at exactly1:00 pm. In hindsight, this should have been a clear indication that war was at hand. The message took some time to decode, and it was not in the hands of the chief of naval operations until about 10:30. General George C. Marshall, the U. S. Army Chief of Staff, did not see the dispatch until about 11:15 am. Chief of Naval Operations, Admiral Harold Stark, despite advice from underlings, did not think that the communication called for any additional instructions to Admiral Kimmel. For his part, General Marshall did decide to issue a new warning to the military

[49]Goldstein, Donald M. and Dillon, Katherine V., eds., *The Pearl Harbor Papers: Inside the Japanese Plans*, Brassey's, 2000.

command and ordered it to communicate with the Navy. The warning did not reach Hawaii until after the attack had begun.

There were a few incidents at Pearl Harbor that by hindsight could be interpreted as a brief warning. Four hours before the attack began, a minesweeper, *U.S.S. Condor*, reported it had sighted a Japanese submarine within the American security zone. About two and a half hours later, the commander of the destroyer *U.S.S. Ward* sent a message saying that he "had attacked, fired upon, and dropped depth charges upon submarine operating in defensive sea area" near Pearl Harbor. While Admiral Kimmel waited for confirmation of this report, the Japanese opened hostilities. About the same time, an army unit, practicing on the radar set after its normal closing time, noticed a large flight of planes on the screen. When he telephoned the command center, the minor officer in charge told him to disregard the observation, because a flight of B-17 bombers from the United States was expected at that time.[50]

The first Japanese dive-bombers flew over Pearl Harbor at 7:55 a.m. local time. The first wave of nearly two hundred aircraft included torpedo planes and bombers, accompanied by a few fighters. Various army airfields came under savage attack. Because General Short was most concerned with the possibility of local sabotage, and with Admiral Kimmel's concurrence, U. S. military aircraft were packed tightly together at the Naval Air Station on Ford Island and adjoining Wheeler and Hickam fields. Most U. S. aircraft were destroyed on the ground by Japanese strafing. At Wheeler Field in particular, the destruction was complete. Of the 126 planes on the ground, 42 were totally destroyed, 41 were damaged, and only 43 were left fit for service. Only six U.S. planes got into the air to repel the attackers of this first assault. In total, more than 180 U. S. aircraft were destroyed.[51]

Of course, the real object of the attack was the American fleet and the Japanese aircraft directed a massive action against Admiral Kimmel's fleet. Ships anchored in the harbor were perfect targets for the Japanese

---

[50]Zimm, Alan D., *Attack on Pearl Harbor: Strategy, Combat, Myths, Deceptions*, Havertown, Pennsylvania: Casemate Publishers, 2011.

[51]Borch, Frederic L.; Martinez, Daniel, Kimmel, *Short, and Pearl Harbor: The Final Report Revealed,* Naval Institute Press, 2005.

bombers. Additionally, because it was Sunday morning, a time the Japanese cleverly chose, the ships were not fully manned. Most of the damage to the battleships was inflicted in the first 30 minutes of the assault. The battleship *U.S.S. Arizona* blew up with a tremendous explosion. Riddled with bombs and torpedoes, the *U.S.S. West Virginia* settled on an even keel on the bottom of the harbor. The *U.S.S. Oklahoma*, hit by four torpedoes within five minutes, rolled completely over, with its bottom and propeller rising above the waters of the harbor. The *U.S.S. California*, the flagship of the Pacific Battle Force, was abandoned as it slowly sank in shallow water after being hit by torpedoes. The *U.S.S. Utah* also was sunk. It had been commissioned as a battleship, but was now considered a target ship. Few of the moored vessel escaped damage.

At about 8:50 am the second wave of the attack began. Although somewhat less successful than the first wave, it still inflicted heavy damage. The first wave aircraft hit the battleship *U.S.S. Nevada* with a torpedo, but its position at the end of Battleship Row allowed it greater freedom of action than the other capital ships moored nearby. As it attempting to get underway when the second wave hit, it was struck by seven or eight bombs and was grounded near the entrance of the channel. The battleship *U.S.S. Pennsylvania* was set afire by bombs and became a blazing wreck. Two destroyers moored near it were utterly destroyed. Rocked by a great internal explosion the destroyer *U.S.S. Shaw* was split in two. Shortly after 9:00 am the Japanese, their bombs and torpedoes launched, withdrew.

Other ships suffered damage. Among cruisers was the *Helena*, hit by one torpedo but returned to service January 1942. The *Raleigh* was hit by one torpedo but returned to service February 1942. The *Honolulu* experienced a near miss, sustained light damage and remained in service. Destroyers. suffered less. The *Cassin* was in dry-dock with the *Downes*. It was hit by one bomb, burned, but returned to service February 1944. The *Downes* caught fire from the *Cassin* and burned, but returned to service November 1943. The *Helm* was underway to West Loch, but damaged by two near-miss bombs. It continued on patrol, was dry-docked on 15 January 1942 and sailed again on 20 January 1942. The Shaw was hit by three bombs, damaged, but returned to service in June 1942. Among the auxiliaries *Oglala*, a mine-layer; the *Vestal*, a repair ship; *Curtiss*, a seaplane tender; *Sotoyomo*, a harbor tug; a floating dock were all damaged, but all returned to service.[52]

Was the Pearl Harbor attack a success? The *Arizona* and the *Oklahoma* were destroyed with great loss of life, and six other battleships suffered varying degrees of damage. Three cruisers, three destroyers, and other vessels were also damaged. U.S. military casualties totaled more than 3,400, including more than 2,300 killed. Heavy damage was inflicted on both army and navy aircraft on the ground. However, of the battleships only the *Arizona* and *Utah* were damaged beyond repair and *Utah* was considered obsolescent if not obsolete. *Oklahoma* was raised, turned back upright, towed to dry dock, and salvageable parts removed, but was not used in combat. All the other battleships re-entered the war before its conclusion. Other smaller vessels were either repaired or replaced.

However, World War II was the war of the aircraft carrier. It would appear that battleships did not make much of an impression during the entirety of the war with perhaps a very few exceptions. The Japanese claimed the unannounced attack, made without a declaration of war, a great victory. This was the first part taken in what has been universally called the Victory Disease.

The tally was impressive but not sufficient to remove the United States from an active role in the Pacific. Nagumo had sunk five American battleships and severely damaged three others. He also sank or severely damaged eleven other vessels, including cruisers, destroyers, and several auxiliaries. Nagumo considered the cost of pressing his luck and assuring full victory and found the following excuses to withdraw. He seemed to be far more interested in returning his fleet intact than in winning a great victory.

While the Japanese lost only 29 aircraft, 74 other aircraft were damaged by anti-aircraft fire from the ground and thus removed from service. The damaged aircraft disproportionately included dive and torpedo bombers. The defensive fire was sufficiently effective to prevent many Japanese aircraft from truly fulfilling their missions. The loss of so many attack aircraft after the first two waves was a harbinger of future

---

[52]Hoyt, Edwin P., *Pearl Harbor*, G. K. Hall, 2000.

disaster at which time American forces would indeed be fully prepared. Nagumo ignored the fact that his losses were far less than the high command had calculated and thus larger losses subsequently would have balanced things. The Japanese lost all five midget submarines, perhaps one or two fleet submarines, and fewer than 100 men. The Japanese task force retired from the theater of battle without being attacked.

Nagumo had absolutely no idea where the American carriers might be. He calculated that he ran considerable risk to his own forces should they be located first and attacked while his aircraft were out searching. He assumed the worst scenario for his own forces, that the American carriers would indeed seek him out and that the aircraft could wreak considerable damage.

Whether his intelligence was faulty or he miscalculated, Nagumo apparently assumed that the obsolescent aircraft on American carriers could indeed inflict serious damage. American torpedo aircraft especially were not competitive, as they proved later at Midway. And American torpedoes were unreliable at the early stages of the war.

He also figured that insufficient daylight remained after recovering the aircraft from the first two waves for the carriers to launch and recover a third flight before dark. This decision stands in stark contrast with the decision the U.S. Navy made at a later date.

Moreover, the destroyers accompanying Nagumo lacked the fuel necessary to remain on duty much longer. This is the most lame excuse of all as it involves a logistic challenge that should have been considered long since.

Much has been made of Yamamoto's hindsight, but, in keeping with Japanese military tradition not to criticize the commander on the spot, he did not punish Nagumo for his withdrawal. By all reasonable standards the operation was almost certainly going to be one of a kind. No Japanese fleet could realistically be expected to travel so far and come to Hawaii again. Its mission was to seek out and destroy the U. S. carriers, absent from the harbor, and bombard and destroy various strategically important facilities on Oahu. What was not accomplished in this attack would not be rectified by later missions.[53]

---

[53]Prange, Gordon William; Goldstein, Donald M. and Dillon, Katherine V. D*ecember*

Months later Yamamoto focused on Nagumo's failure to seize the initiative to return and destroy the dry docks, fuel installations, and repair facilities. Yamamoto also lamented Nagumo's failure to seize the initiative to seek out and destroy the American carriers which were absent from the harbor.[54]

When we evaluate the results of the attack on Pearl Harbor, we must remember that the very first bombs did not do enough. The four American carriers harbored there were the primary target of the Imperial Japanese Navy and they were not present. When IJN abandoned launching the third wave of Japanese aircraft, it also spared other notable targets at Pearl Harbor such as its oil depots. As, as is always to be expected in such large-scale battles, the IJN bombers also missed a few of the most important targets.

Those four US aircraft carriers stationed at Pearl would prove to be invaluable to the decimated American Navy in 1942. After all, this was not a war that required battleships, but one which relied heavily on aircraft and carriers. These craft had the far greater range and could also provide greater support for land battles. Had these carriers been at Pearl Harbor, they might have been lost, and this would have provided Japan with an enormously greater advantage in 1942. As it was, the Allies won the aircraft carrier naval battles that did emerge in 1942, precisely because those carriers had remained unscathed. At the Battle of Coral Sea, and then the Battle of Midway, the Japanese carrier fleet would be decimated by those that had survived at Pearl Harbor. Hundreds of Japanese aircraft were lost, as well as five aircraft carriers, and the strategic targets of Port Moresby and Midway Island remained with the Allies. Much of this was done at the decisive Battle of Midway, which is the battle that some say had the most impact in the war. Had the American carriers been wiped out at Pearl Harbor, then Japan might have easily won such battles.

7, 1941: The Day the Japanese Attacked Pearl Harbor. McGraw-Hill, 1988. See also Conroy, Hilary, and Harry Wray, Eds., *Pearl Harbor Reexamined: Prologue to the Pacific War. Honolulu: University of Hawaii Press, 1980.*

[54]See Smith, Carl, *Pearl Harbor 1941: The Day of Infamy*; Osprey Campaign Series #62, Osprey Publishing, 1999; and Stille, Mark E., *Tora! Tora! Tora!: Pearl Harbor 1941*; Osprey Raid Series #26, Osprey Publishing, 2011.

Battle of Midway

At precisely 10.26 a.m. on 4 June 1942, the course of World War Two in the Pacific changed entirely, for at that moment thirty-seven Douglas Dauntless bombers from the *U.S.S. Enterprise* dived on two Japanese aircraft carriers. Within minutes both ships were ablaze, as they were rocked by the explosion of fuel lines, badly stored explosive ordnance and gasoline tanks. Within the next six hours the other two carriers in the Japanese fleet were also destroyed. The Imperial Japanese Navy (IJN), which had dominated the Pacific for the first six months of the Great Pacific War, was now in ruins, effectively ending Japanese hopes of an empire. The Battle of Midway was among the most rare of military engagements: it was a truly and completely decisive victory.[55]

Following their sneak attack at Pearl Harbor in an undeclared war in December 1941, the IJN's six large aircraft carriers, operating over 400 aircraft, had roamed about the Pacific unimpeded. IJN had sunk two major British warships in the Indian Ocean and bombed the Port of Darwin, Australia. IJN was minimally halted in early May 1942 at the Battle of Coral Sea. One carrier was sunk and another damaged in an inconclusive battle with American carriers. IJN had chosen to withdraw, refit, regroup and try again, in spite of the fact that it had more than sufficient forces to continue. It also ignored the fact that time was on the side of the United States as it began to change over its industrial might from civilian to military enterprises.[56]

The American Doolittle raid of 18 April 1942, largely a

---

[55]We have followed closely the work by Symonds, Craig L. *The Battle of Midway*. Oxford University Press, 2011. Solidly based on the latest historical understandings and written with an entertaining style, this is the best general history of the battle. It is particularly strong in its coverage of the background and the events leading up to the pivotal battle.

[56]Hoyt, Edwin P. *Blue Skies and Blood: The Battle of the Coral Sea*. New York: Pinnacle Books, 1975.

propaganda air attack on Tokyo launched from the carrier *U.S.S. Hornet,* prompted the Japanese to prepare Alfred Thackeray Mahan's great final battle with the United States. The quasi-military and IJN high command ordered Admiral Isoroku Yamamoto to plan this final showdown with the remnants of the American fleet.[57] Doolittle's raid had insulted Japan and theoretically threatened the life of the emperor.

Confident that he had considerable advantage in terms of both numbers and quality of ships, Yamamoto planned to destroy the American carrier fleet. His plan was typically but unnecessarily complex and confusing even to his own men. He planned a diversionary attack on the Alaskan coast, which he was confident would draw the American Navy north to protect its own native soil. Only then would he launch his main attack on Midway Island on the following day. Realizing they had been fooled, the American Navy would turn about and head south, sailing directly into IJN's ambush. The focal point of his plan, the aptly named Midway Island, was America's last remaining base before reaching the Hawaiian Islands. It had long been used as an Allied air base for it was halfway across the Pacific, some 1,200 miles) west of Hawaii.[58]

Yamamoto's opposite number, American Admiral Chester Nimitz, commanding from Pearl Harbor following the replacement of Admiral Husband Kimmel, had fewer ships, but he had gained a priceless advantage when his intelligence service succeeded in deciphering IJN's radio code. Since the Japanese fleet was widely dispersed, Yamamoto transmitted his plan by radio. American code breakers were able to recover about 90 per cent of the plan. Intelligence was certain that the object was Midway, based on an earlier intercept referring to Midway as AF. A simple trick, reporting an alleged salt desalination failure, yielded the necessary confirmation. A Japanese intercept noted that Midway's fresh water

---

[57]Parshall, J., and Tully, A. *Shattered Sword: The Untold Story of the Battle of Midway.* Washington, D.C.: Potomac Books, 2006. This book is a recent reexamination of the Battle of Midway from the Japanese perspective and is likely to become the definitive work on the Japanese side of Midway.

[58]Barde, Robert Elmer, "The Battle of Midway: A Study in Command" Ph.D. dissertation, University of Maryland, 1971. See also Barker, A. J., Midway*: The Turning Point.* New York: Ballantine Books, 1971.

condenser was malfunctioning. AF was Midway.[59]

We say a second time that Yamamoto's plan was needlessly complex. Although he had six carriers, and four should be sufficient for the Midway operation, there was no real need to allocate two carriers to an Alaskan diversion. Yamamoto also relied upon an air reconnaissance to locate the American carriers at Pearl Harbor. Submarines would refuel the spy planes. When the carriers moved from Hawaii a submarine ambush was supposed to sink them. As it was, the submarine reconnaissance was thwarted by an American warship, and the submarine ambush arrived too late to ambush the American fleet.

The usefulness of the Alaskan diversion depended wholly upon the American willingness to send its ships to Alaskan waters. As with so many Japanese plans, this one depended on the Americans doing exactly what the Japanese wanted them to do. Yamamoto's plan gave no thought to the United States ignoring the diversionary attack. This phase of Yamamoto's plan proved to be a waste of resources since the Americans failed to respond to it. The two carriers sent toward Alaska would have proved to be far more useful in the Midway operation.[60]

Japanese planners made a second grievous error. They worked on the assumption that the Americans had no more than two carriers, for they were confident that IJN had sunk the *U.S.S. Yorktown* at the Battle of the Coral Sea. In the actual event, the damaged *Yorktown* was repaired and made battle-worthy in just two days at Pearl Harbor. Even so, there remained the massive imbalance between the two fleets. The Americans had three carriers, eight cruisers and 15 destroyers divided into two task forces. Yamamoto commanded four large and two medium carriers, eleven battleships, along with numerous cruisers and destroyers. Additionally, there was Admiral Kondo's amphibious task force with which he planned to occupy Midway.

As was typical of Japanese naval planning, Yamamoto divided his force into three main groups. There were four big carriers, the battle-fleet,

---

[59]Holmes, W., *Double-Edged Secrets: U.S. Naval Intelligence Operations in the Pacific During World War II*. Annapolis, Maryland: Naval Institute Press, 1979.

[60]Stille, Mark. *USN Carriers vs IJN Carriers: The Pacific 1942*. New York: Osprey, 2007.

and the invasion force. These three groups were too far apart for mutual support. The Japanese carrier group operated in close order, commanded by Admiral Nagumo, who had led them for the attack at Pearl Harbor. None of the groups was wholly self-supporting.[61]

The United States could not possibly win a direct engagement. It staked everything on being able to exploit the intelligence coup, and ambush the enemy. He quietly reinforced the air units on Midway, using the island as an unsinkable aircraft carrier. Nimitz ordered his carriers to take positions to the north-east of the island, and wait in ambush for the Japanese carriers to appear.

The American tactics relied on the singular and unique characteristics of carrier warfare. American commanders knew the first attack would be decisive for either side, since the carriers would be full of fuel and explosive ordnance, and hence highly vulnerable to enemy bombs and torpedoes. The American tactical commanders were Admiral Frank Fletcher and Admiral Raymond Spruance, with Admiral William Halsey hospitalized. Both admirals knew they were playing for very high stakes and thus kept the two task forces separate.

On 3 June, American bombers from Midway located and attacked the Japanese invasion fleet as it approached the island. Large bombers dropping ordnance from high levels rarely did much damage to ships and that was the case here. The next morning Nagumo launched about one hundred planes against the island. U. S. Marines had about fifty planes, but most were obsolescent. The old Vindicators were called "wind indicators." The planes from Midway attempted to defend but had no success. The Japanese strike was initially effective, although Midway remained operational.

Now the fundamental flaw in the Japanese plan became apparent. The carriers were expected to strike Midway and engage the expected two (not three) American carriers when they eventually arrived. Yamamoto's plan was predicated on the arrival of his fleet well in advance of the arrival of the American carriers. It also assumed that resistance on Midway would be reduced before the American carriers arrived. The IJN plan absolutely

---

[61]Fuchida, Mitsuo and Masatake Okumiya. *Midway: The Battle that Doomed Japan, the Japanese Navy's Story.* Annapolis, Maryland: Naval Institute Press, 1955.

had not expected the American fleet to be in position before its forces arrived. We must reiterate that whatever merit Yamamoto's plan had, its worth was substantially reduced if the United States did not behave as it anticipated.

Yamamoto's scouts had reported the presence of hostile warships, but they had not seen any American carriers. Reports of the carriers arrived only after the attack on Midway had begun. The scout plane which had been detailed to cover the sector where the American ships were waiting had been delayed. Now there were two critical tasks at hand-- seizing the island and attacking the American fleet. Which should he attack first? Nagumo hesitated when he should have acted. This indecision turned the battle.

For their part, Spruance and Fletcher had only one task, and that was to find and destroy the Japanese carriers. Once the IJN carriers had been located, Spruance launched a full strike from the *Enterprise* and the *Hornet*. Fletcher initially hesitated, but then decided to launch half his strike force.

Meanwhile, Nagumo had decided to proceed with a second strike on Midway since his returning pilots reported that Midway's airstrips were still operational. The report of the American carrier *Yorktown* changed Nagumo's mind. He ordered his aircraft's armament to be changed back to torpedoes. Additionally, he had to recover planes that had bombed Midway, and not allow his pilots to ditch in the ocean. The planes he planned to send against the Enterprise had to be accompanied by righting aircraft so he had to refuel his fighters. It was a careful response, typical of an old-school officer like Nagumo. But it was proved to be the wrong response.

American commanders had planned to attack the Japanese force in a combined assault, using both low-level torpedo bombers with high-level dive bombers. Both planes were to be escorted by fighters. All planes were supposed to attack at the same time in order to overwhelm the defenders. As it was, the various squadrons became separated, and some of *Hornet*'s inexperienced units even missed the battle entirely.

The torpedo bombers were the first to reach the Japanese carriers. Flying in at low level, they were badly mauled by fast, maneuverable

Japanese Zero fighters. Between 9.30 and 10.24 a.m., 47 out of 51 American torpedo planes were shot down without inflicting any damage upon the enemy ships. Initially, he weathered numerous attacks and emerged unscathed, while delivering a heavy blow to the Midway air base. Now it was his turn as he would launch a huge strike and finish off the American carriers.[62]

In the actual case, it was the Americans who had all the luck that day. While the dive-bombers from the *Enterprise* were lost at first, by good luck they located their target. They spotted and then followed a Japanese destroyer which was steaming at high speed to rejoin the carriers after driving off an American submarine.

The dive-bombers arrived over the Japanese carriers just as the last American torpedo planes were being shot down. The Zero fighters were flying near sea level and consequently out of position and unable to counter *Enterprise*'s dive-bombers. These pilots carried out a textbook perfect dive-bombing attack on the big Japanese ships. The Japanese had been caught unprepared partly because they did not have radar to spot the threat. Within five minutes the dive bombers hit the carriers *Kaga* and *Akagi*, igniting uncontrollable fires. Their decks were cluttered with bombs, torpedoes and hoses charged with high octane fuel, while every plane on deck was also a huge gasoline bomb waiting to explode. It took only a few thousand-pound bombs to set these massive ships ablaze.

Yorktown's dive-bombers followed, attacking the carrier *Soryu*. Only the carrier *Hiryu* escaped, hidden by a rain squall. Within minutes Nagumo had ordered the *Hiryu* to launch a hastily assembled half-strength strike-force consisting of forty planes. They headed for *Yorktown*, which was the only American carrier the Japanese actually located during this battle.

With radar warnings Yorktown took all possible precautions. The crew purged its fuel lines and filled them with carbon dioxide. They launched their fighter defenses, and readied anti-aircraft guns. But *Yorktown* was torpedoed twice and also hit by three bombs. Although seriously damaged, Yorktown did not catch fire and remained afloat. The

---

[62]Kernan, Alvin, *The Unknown Battle of Midway*. New Haven, Connecticut: Yale University Press, 2005.

United States Navy had invested heavily in damage control, fire-fighting and safety systems and these proved to be very effective. The Navy had also learned a costly lesson at the Battle of the Coral Sea, where a carrier exploded. *Yorktown* was temporarily saved, but its planes were forced to land on the other carriers. By launching its aircraft *Hiryu* had revealed itself. Within six hours American aircraft caught and destroyed the *Hiryu.*

Upon learning of the disaster that had befallen his fleet, Admiral Yamamoto, actually located hundreds of miles to the west with the main battlegroup, reversed course.[63]

Admiral Spruance was cautious, and, fearing a trap, did not give chase. On 6 June American aircraft located and sank a Japanese heavy cruiser. The following day a Japanese submarine caught up with the damaged *Yorktown* and sank it with a torpedo.

The Japanese lost four carriers, a heavy cruiser, 3,500 men and 270 aircraft during the battle. Many highly trained aircraft maintenance teams went down with the four carriers. Most experts conclude that the pride of Japanese Naval aviators was lost and the IJN never quite recovered from the loss of so many first line pilots.

The United States lost one carrier, about a hundred men and 130 aircraft, many obsolescent if not obsolete. In return they had halted the mighty Imperial Japanese Navy, and gained the initiative. Were a modern historian to update Creasy's *Decisive Battles of History*, Midway would merit inclusion. The tide of war had turned inexorably in favor of the United States.[64]

Once again, the Japanese failed to exploit, let alone win, the Great Decisive Battle, following their interpretation of Mahan. Yamamoto could have recalled the two carriers from the ill-fated Alaskan adventure and pressed on. Kondo might still have invaded because he had a mighty force of battleships and cruisers at his disposal. The large craft could have moved quite close to Midway and shelled it into oblivion in much the same way the IJN did at Wake Island. There was precious little the Marine

---

[63]Isom, Dallas Woodbury. *Midway Inquest: Why the Japanese Lost the Battle of Midway.* Bloomington, Indiana: Indiana University Press, 2007.

[64]Mitsuo, Fuchida, Mitsuo and Okumiya, Masatake. *Midway: The Battle that Doomed Japan.* Hutchinson, 1957; Prange, Gordon, *Miracle at Midway.* McGraw-Hill, 1982; and Lundstrom, John. *The First Team.* U. S. Naval Institute Press, 1990.

garrison at Midway could have accomplished.[65]

In his magnificent story of the Battle of Midway, University of Maryland Professor Gordon Prange[66] made the point well: Yamamoto never made clear whether the first priority was destruction of the American fleet or the occupation of the island of Midway. Professor Prange wrote, "From the very beginning, Operation MI was a monster with two heads, each arguing with the other."

Faced with two distinct choices, Admiral Nagumo had no clear orders. If both objectives were of equal value and importance why divide IJN's forces? This dissipation of strength violates one of Admiral Mahan's basic rules. IJN made the wrong choice. It decided to concentrate on the capture of Midway.

The decision to add the Aleutian campaign to IJN's plan made no sense. Whether it was true that by attacking the Aleutians IJN hoped to lure the American fleet from the central Pacific or not, it was an unnecessary and expensive division of forces. The decision to split the forces available, sending one to a distant and unimportant target violated the cardinal rule of establishing and maintaining superiority at the point of contact. IJN's numerical superiority could have been absolutely overwhelming had it arranged its forces according to teaching. Incredibly, Yamamoto, having assembled the greatest fleet to sail in any ocean to date, blunted its advantage by dispersing it.

The invasion of Midway, as with any similar operation, mandated that the invading force create and maintain a firm schedule. Meeting a powerful enemy in mid-ocean required great flexibility. There was no question that defeat of the American fleet was of paramount importance and had to be the number one priority without any question whatsoever. The Allies had few capital ships remaining although the fabrication of more would eventually follow. IJN's opportunity to achieve absolute naval supremacy in the Pacific fit into a clearly limited window in time. Following the defeat of the American naval force,

---

[65]Smith, Peter C. *Midway Dauntless Victory; Fresh perspectives on America's Seminal Naval Victory of 1942*. Barnsley, UK: Pen & Sword, 2007.

[66]Prange, Gordon W. *Miracle at Midway.* McGraw-Hill, 1982; also, Penguin 1983. See especially pp. 376-82.

Japan could occupy Midway Island at will.

Professor Prange reminds us that the invasion of Midway was essentially defensive, designed to secure the vital outer defensive ring which was to protect Japan and its newly-conquered empire. Midway was not an offensive move, designed to strengthen Japan. IJN did not consider whether the Midway campaign would further its national interests, or what the consequences would be if Japan was defeated. Precisely what would Japan gain by the Midway campaign that would justify its anticipated cost in men and materials, without even considering some the cost were there a catastrophic defeat?

The fact that IJN determined that it should have four distinct fleets also hindered the chances of victory. IJN determined that each fleet be specialized. Logic dictates that each fleet should be complete and self-sustaining. As it was the several bodies did not coordinate. One fleet could not assist another. Admiral Yamamoto accompanied his fleets and was thus precluded from directing movements for to have done so would have violated radio silence.[67]

In part due to the infection of the Victory Disease, the Japanese failed to maintain secrecy as they had done before the sneak attack on Pearl Harbor. Too many people knew, many needlessly. IJN did not suspect that Americans could possibly crack its naval code so was late in making changes that might have maintained secrecy. IJN allowed small, but preventable, failures to undermine its plan. The submarine monitoring line was so late in its deployment that the American Navy had already passed. Likewise, the failure to scout the American Navy with seaplanes precluded knowledge of the movement of the American aircraft carriers.

The Battle of Midway was a catastrophic defeat for the Japanese, suffered at the hands of an inferior American force. It was a disaster from which the IJN would never fully recover. The sinking of four fleet carriers, a heavy cruiser, and loss of nearly 250 aircraft manned by veteran pilots could never be made good by the Japanese industrial base or her aviation training program. U. S. losses, the carrier *Yorktown*, the destroyer

[67]The Japanese side of the Battle of Midway is told in Parshall, Jonathan and Tully, Anthony. *Shattered Sword, The Untold Story of the Battle of Midway.* Washington, DC: Potomac Books, 2005.

Hammann, and 100 aircraft, were serious but temporary, and the torpedo aircraft lost were obsolescent and thus their absence was of little consequence. Destroyed American aircraft were replaced by superior models. Capital ship loss was offset once *Saratoga* was repaired and returned to action, and the new commissioned *Buchanan* joined the Pacific Fleet in mid-June 1942.

Even the god-emperor Hirohito and his inner circle were not told of the complete truth until several days after the battle. At an IJN meeting on the 10th, the extent of the loss was concealed from the army (IJA) and a cabinet minister was appointed to oversee control of the information. What remained of the Combined Fleet did not return in full until June 14 and was partially concealed by poor weather and fog. With the concurrence of Hirohito, Naval General Staff then ordered that the wounded would be treated with the utmost secrecy. Survivors returned to service as quickly as possible. They were to have little to no contact to anyone, including family. This plan included placing injured survivors in special security-controlled wards for months. Those able to serve again were sent to either backwater posts or fed into the meat grinder of the South Pacific once fighting started in the Solomon Islands. A civilian cameraman who had been aboard *Akagi* was held and warned not to return to Tokyo before being shipped out. The aftermath of the Battle of Midway represented the first time the Japanese government fabricated news reports. Heretofore, Japanese authorities had providing filtered but optimistic views of the war. The official IJN reports lied, saying that the IJN had suffered the loss of two carriers, but had also destroyed two U. S. Navy aircraft carriers and also had inflicted considerable damage to other forces. Even the Navy was fed misinformation. The *Kaga* and *Soryu* were removed from the ship registry list, but *Akagi* and *Hiryu* remained on the active list. However, no men assigned to them and they were said to be damaged. Journalists who knew the truth were restrained.[68]

The United States Navy was expanding with lightning speed and by August 1942, the following aircraft carriers were completed or were awaiting completion in American shipyards: six fleet aircraft carriers;

---

[68]Smith, Douglas V. *Carrier Battles: Command Decision in Harm's Way*. Annapolis, Maryland: Naval Institute Press, 2006.

eight light aircraft carriers; and twenty-two escort aircraft carriers. Several of the damaged battleships at Pearl Harbor were undergoing repairs. Additionally, American naval yards were completing five new battleships, four heavy cruisers, nine light cruisers, 113 destroyers, and 29 submarines. In various stages of construction were seventeen *Essex* class aircraft carriers.

Naval Action at Guadalcanal

As we have seen the U. S. Navy made its first impact against the Japanese advance in the Battle of the Coral Sea and, more significantly, the Battle of Midway. The next U.S. military move involved a two-pronged assault in the Solomon Islands and New Guinea. The first major land attack launched by the Allies against Japan in World War II took place on Guadalcanal in a battle that lasted from 7 August 1942 until 9 February 1943.

Japan invaded Guadalcanal, located in the Solomon Islands, in July 1942 and quickly became one of the most important Japanese strongholds due largely to its proximity to Australia. Soon, the Japanese began constructing a full-fledged airbase on the nearby island of Guadalcanal. Allied intelligence had deciphered messages that showed Japanese intentions in the area even before the first Japanese construction units landed. America responded by dispatching the 1st Marine Division. For the next four months, the waters around Guadalcanal would become the scene of the most bitter and protracted naval struggle of the Pacific War. Although neither party realized it at the time, the Japanese had found their Great Decisive Battle which their Mahan-inspired naval doctrine called for. As events unfolded, it became not just a battle, but an entire campaign. It was one the Japanese could not win, partially because of the material superiority of their adversary, and partially because of the willingness of the Americans to accept stunning naval losses in order to hold the island. This battle gradually eroded the vital cruiser and destroyer strengths of the Japanese Navy. The Japanese built an airfield at Lunga Point and placed artillery positions in the hills nearby. By the end of the summer 1942 Japan had about 8,400 men on the island by August.

The Allies also intended to use Guadalcanal and Tulagi as bases in supporting a campaign to eventually capture or neutralize the major Japanese base at Rabaul on New Britain. On 7 August 1942, U.S. Marines landed on the northern beaches of Guadalcanal, moving under the umbrella of Naval gunfire ahead of them. The Japanese defenders, who had occupied those islands since May 1942, were outnumbered and overwhelmed by the Allies, who captured Tulagi and the airfield which was under construction on Guadalcanal, renaming it Henderson Field in honor of a Marine pilot. Surprised by the Allied offensive, the Japanese made several attempts between August and November to retake Henderson Field.[69]

Over the next three months, the Marines secured the airfield and a 6-mile-wide section of the beach. On 13 October 13, an Army unit arrived to reinforce the Marines. The combined American military strength repelled a Japanese attack on the 23rd, inflicting heavy losses on the Japanese and pushing them out further during the rest of the month. On 4 November Army infantry fought a new force of some 1,500 Japanese troops that had landed on the beach at Koli Point. They killed half the Japanese force, while the rest escaped into the jungle.[70]

Three major land battles, seven large naval battles marked the Guadalcanal campaign. Five were night-time surface actions at which the Japanese at that time excelled. An additional two engagements involved air-carriers. And there were almost daily battles for air supremacy. Japanese aircraft from Rabaul made constant sorties against the Marines on Guadalcanal.[71]

One of the most important engagements took place on the night of 8 to 9 August 1942. Known as the Battle of Savo Island, it pitted Allied cruisers and destroyers against a strong force of the Imperial Japanese Navy. As the transports unloaded on the night of 8–9 August, two groups of screening Allied cruisers and destroyers, under the command of British Rear Admiral Victor Crutchley were surprised and defeated by a Japanese force of seven cruisers and one destroyer. The Japanese fleet was part of the 8th Fleet based at Rabaul and Kavieng and commanded by Japanese

---

[69]Hersey, John. *Into the Valley: Marines at Guadalcanal.* Lincoln: University of Nebraska Press, 2002.

[70]Coggins, Jack, The Campaign for Guadalcanal: A Battle That Made History. Garden City, NY: Doubleday and Co., 1972.

[71]Leckie, Robert, *Challenge for the Pacific: The Bloody Six-Month Battle of Guadalcanal.* New York: Da Capo Press, 1999.

Vice Admiral Gunichi Mikawa. In the Battle of Savo Island IJN sank one Australian and three American cruisers. Additionally, IJN severely damaged one American cruiser and two destroyers. The Japanese suffered only moderate damage to one cruiser.[72]

On 7 August, 1942, the U.S. 1st Marine Division landed on the islands of Guadalcanal and Tulagi and took control of the uncompleted airfield. Vice Admiral Gunichi Mikawa, the newly installed commander of the Eighth Fleet, working from the major base of Rabaul, assembled every warship at his disposal, arriving off of the southern shore of Savo Island early in the morning of 9 August. Ahead of him were several groups of Allied warships, with their numbers divided into three patrolling squadrons. Beyond the Allied warships lay a transport anchorage off of Lunga Point.

Mikawa's ships slipped unseen past the destroyer pickets at the mouth of the Sound, and soon came upon the southern group of Allied warships which consisted of two heavy cruisers, Australian *HMAS Canberra* and *USS Chicago* and two US destroyers. Mikawa's force first launched torpedoes and then followed up with devastating salvos of 8- and 6-inch gunfire. *Canberra* was sinking before she was aware that a battle had been joined. *Chicago* fared only slightly better, but never got into action, and never alerted the Northern Force of Mikawa's attacking squadron. Fifteen minutes later, curving northward around Savo Island's eastern shore, the Japanese came upon the Northern Force, still steaming sedately along in a box patrol pattern. Mikawa's forces had become divided in the earlier exchange, and by pure chance enveloped the Allied Northern force. Caught in a devastating crossfire, Northern Force's three heavy cruisers, *Vincennes, Quincy,* and Astoria, were sunk.

At this point, having slaughtered the Allied escorts, the transport anchorage behind him lay open for Mikawa's taking. But the Japanese admiral's position was not as advantageous as he would have wished. He had no idea that the U. S. carriers under Admiral Fletcher covering the invasion had been withdrawn from the general vicinity. Attacking the

---

[72]Poor, Henry V., Henry A. Mustin and Colin G. Jameson. *The Battles of Cape Esperance, 11 October 1942 and Santa Cruz Islands, October 26, 1942.* Washington, DC: Naval Historical Center, 1994.

transport anchorage would require his slowing, reassembling his scattered squadron, coming about, finding the anchorage, and then attacking it. With essentially no opposition from a surface fleet left, Japan had only to face the American carriers. Mikawa was unaware U. S. Admiral Jack Fletcher was preparing to withdraw with the U. S. carriers, and Japanese intelligence and scout planes failed to make this determination so that Mikawa could capitalize on this situation. Instead of following up on his great victory, Mikawa immediately retired to Rabaul without attempting to attack the transports. Mikawa was concerned about Allied carrier and land-based air attacks should he remain in the Guadalcanal area. It was a spectacular tactical victory for the Japanese, but it was also shorn of the strategic advantage it might have achieved.[73]

With Admiral Fletcher's carriers moving away and thus bereft of his carrier air cover, Admiral Turner decided to withdraw his remaining naval forces by the evening of 9 August. By so doing, he left the Marines ashore without much of their heavy equipment, provisions, ammunition, medical supplies and other logistical support. Too there were some troops still aboard the transports. [74]

Admiral Mikawa's decision not to attempt to destroy the Allied transport ships when he had the opportunity proved to be a crucial strategic mistake. Once again Japanese cowardice grasped defeat from the jaws of victory. Once again, the indecisive Japanese high command was more concerned with returning their ships. With United States still reeling from substantial losses here was Japan's great opportunity to turn back the Allied assault before it really began.[75]

Battle of Cape Esperance

The night of October 11, 1942, Rear Admiral Norman Scott headed a task force which was standing off the entrance to Ironbottom

---

[73]http://www.combinedfleet.com/battles/Guadalcanal_Campaign

[74]Twining, Merrill B. *No Bended Knee: The Battle for Guadalcanal.* Novato, CA.: Presidio, 1996.

[75]The definitive study of these campaigns was compiled by Admiral and Professor Samuel Eliot Morison of Harvard, in Volume IV of his *History of United States Naval Operations in World War II* on Coral Sea, Midway and Submarine Actions, May 1942-August 1942. Boston, MA: Little, Brown and Co., 1960.

Sound. His assigned mission was to prevent Japanese bombardment forces from attacking Marines on Guadalcanal. Simultaneously, a Japanese group, commanded by Rear Admiral Aritomo Goto, approached the entrance to Ironbottom Sound. Upon spotting the approaching Japanese, Admiral Scott adopted a simple battle plan. He kept his ships in a straight line-ahead formation, using the destroyers to illuminate targets, and his cruisers to neutralize the opponent with gunfire. His two light cruisers, *Boise* and *Helena*, each sported fifteen 6-inch guns. For reasons that are unclear, Scott's chose as his flagship, the heavy cruiser *San Francisco*, which was equipped with radar that was inferior to the two light cruisers.

Helena detected the approaching Japanese force on radar, but Scott's distrusted of the information he was receiving from *San Francisco*'s set and demurred, thus allowing the range to close to perilous proximity before opening fire. Although the Japanese force was taken largely by surprise, they quickly realized that Scott had crossed their 'T' Their commander Admiral Goto ordered his ships to execute individual turns to port and starboard in order to clear the area. The Allied force hit the Japanese flagship *Aoba* hard early on, and Admiral Goto was mortally wounded in the opening moments. Seeing the *Aoba* in distress, the captain of *Furutaka* swerved to interpose his vessel between the American force and the flagship. It was to prove a great mistake, as she was destroyed by six-inch gunfire. Likewise, the destroyer *Fubuki*, sailing in the van, was also quickly dispatched. The Japanese retreated, with the Americans breaking off the pursuit after a few hours.

The American destroyer *Duncan*, finding herself between the two main forces, had closed the Japanese and attempted to attack with torpedoes. She was heavily damaged in the process, hit by both Japanese and American shells, and eventually sank. *Boise* sustained heavy damage. In the end, there was yet another flawed victory, this time for the Americans. Given their numerical advantage, the element of surprise, and their superior tactical position, they ought to have inflicted heavier casualties on the Japanese. Instead, they had sunk a cruiser and a destroyer, at the cost of destroyer sunk and a very valuable light cruiser badly damaged. More important, this engagement did nothing to dissuade the Americans from their linear, line-ahead tactics, which would have importance in later fights.[76]

First Naval Battle of Guadalcanal

Japanese intelligence was slow in determining that the Americans had far more troops on Guadalcanal than it had counted in its earlier estimates. Strategically, it determined that Henderson Field had to be either reconquered or at least neutralized. Otherwise the Japanese could not control the seas around the island. Consequently, on 11 November, the Japanese assembled a large convoy of merchant vessels, loaded with enough supplies and ammunition sufficient for a month of fighting. To protect the supply convoy, IJN assembled a task force, built around the battleships *Hiei* and *Kirishima.* The battleships were ordered to shell Henderson Field. On the 13 November, the task force entered Ironbottom Sound. Awaiting it was an American force composed of heavy and light cruisers, and destroyers under Rear Admiral Daniel Callaghan. The Allied fleet repeated its earlier mistake of failing to use its superior radar properly. When the Allies spotted the Japanese force on radar, Callaghan was unaware of the composition, speed, and course of the opposing Japanese. His attempt to cross the Japanese 'T' instead placed his ships on a collision course with the enemy. By the time fighting actually began, the range between the leading elements of each force had closed to 1000 yards. The end result was point-blank combat. Each formation passed through each other, bringing about severe damage to both sides, although the American force was more severely injured. The good result was that American air power sank the *Hiei* as she cruised back up The Slot. Neither side was willing to give up control over the Sound. The stage was set for another brutal battle just two nights later.[77]

Second Naval Battle of Guadalcanal

The Second Naval Battle of Guadalcanal took place on the night of 14-15 November 1942. The American cruiser force was severely depleted because of losses in the first battle of Guadalcanal where the light

---

[76]Cook, Charles Orson. *The Battle of Cape Esperance: Encounter at Guadalcanal.* Naval Institute Press, 1992.

[77]Hammel, Eric. *Guadalcanal: The Carrier Battles - Carrier Operations in the Solomons,* August-October 1942. Naval Institute, 1987.

cruiser *Atlanta* was sunk and the cruisers *Portland* and *San Francisco* were severely damaged. The light cruisers *Helena* and *Juneau* were also damaged. To compensate Admiral William Halsey detached the fast battleships *South Dakota* and *Washington*, under the command of Rear Admiral Willis Lee, heretofore part of the carrier *Enterprise*'s screen, to assist in the defense of Ironbottom Sound.

Another Japanese force, led by the *Kirishima*, assisted by the heavy cruisers *Atago* and *Takao,* returned to again try to neutralize Henderson Field. *South Dakota* lost power due to faulty electrical equipment, and was unable to contribute to the battle. The *Kirishima* and several other Japanese ships riddled her and put her radar out of commission. *Washington* approached undetected to within 8,400 yards of the Japanese force and disabled *Kirishima* and sank *Anyanami*. Fearing torpedo attack, Admiral Lee withdrew. In the course of three days of constant fighting in and around Ironbottom Sound, IJN had lost two battleships, one heavy cruiser, three destroyers and eleven combat transports with 5,000 infantrymen.[78]

Battle of Tassafaronga

The Naval Battle of Tassafaronga occurred on the night of 30 November 1942. Japanese planners had made the decision to abandon Guadalcanal. IJN was charged with removing as many troops as possible before capitulation. Destroyers, which previously cruised down at night to drop supplies, now began to evacuate Japanese infantry units. On the night of 30 November, Rear Admiral Raizo Tanaka led a force of eight Japanese destroyers down The Slot. Of the eight, six were heavily loaded with supply drums and had landed their torpedo reloads, drastically reducing their combat effectiveness. Only Tanaka's flagship *Naganami* and *Takanamii* were fully combat ready. Awaiting the Japanese force in Ironbottom sound was a dramatically superior American force, consisting of four heavy cruisers, a light cruiser, and six destroyers. As it was, Tanaka's lookouts detected the U. S. ambush before it was sprung. American linear battle tactics cost them, as the Japanese destroyers

[78]Wheelan, Joseph. *Midnight in the Pacific: Guadalcanal--The World War II Battle That Turned the Tide* of War. DaCapo Press, 2017.

wheeled, fired torpedoes, and departed. The American task force sank *Takanami.* IJN sank U. S. heavy cruiser *Northampton*, and damaged cruisers *Minneapolis, New Orleans* and *Pensacola.* Not only that, but Tanaka had managed to drop off his supply barrels. It was a humiliating defeat for the U. S. Navy. Nonetheless, the Japanese continued their retreat from Guadalcanal. Japan did nothing to follow up on this major victory.[79]

[79]Crenshaw, Russell Sydnor, Jr. *Battle of Tassafaronga.* Naval Institute Press, 2010.

Battle of Philippine Sea

The Battle of the Philippine Sea took place on19 and 20 June 1944. The Japanese stronghold of Turk island had been eliminated as an effective base, and the Japanese Navy driven from their area by the summer of 1944. Now, America could complete its invasions of the Marshall Islands without interference from IJN. United States now focused on the Mariana Islands. Once secured these islands could host long-range American B-29 aircraft with the range necessary to hit the Japanese Home Islands. The Japanese were very well aware of this fact. On 12 June 1944, U.S. carriers started flying a series of air strikes on the Marianas, convincing Admiral Toyoda that the U.S. was preparing to invade. This move surprised the Japanese who expected the next U. S. target to be farther to the south, either the Caroline or the Palatals. Thus, at that time, the Marianas were protected with only a weak force of 50 land-based aircraft.

To counter this move, once again the IJN high command planned yet another Great Decisive Battle designed to prevent their capture. Admiral Jisaburo Ozawa's plan was to use the longer range of his aircraft to good effect by shuttling them between his carriers and the airbases on Guam and Saipan, while keeping his carriers out of range of an American counter-strike.

On 15 June, the American submarine *U.S.S. Flying Fish* sighted a Japanese carrier and battleship force coming out of San Bernardino Strait. An hour later, another submarine, *U.S.S. Seahorse,* spotted a battleship and cruiser force sailing up from the south, some 200 miles east of Mindanao. All American submarines were under orders to report sightings before attempting to attack, so Flying Fish waited until nightfall, then surfaced to radio in its report. Fifth Fleet commander Admiral Raymond Spruance was convinced that a major battle was at hand. After consulting with Admiral Chester Nimitz at Pacific Fleet Headquarters in Hawaii, he ordered Task

Force 58, which had sent two carrier task groups north to intercept aircraft reinforcements from Japan, to reform and move west of Saipan into the Philippine Sea.[80]

Admiral Mitscher realized that there was a chance of a night surface battle with Ozawa's forces. Mitscher's Chief of Staff Arleigh Burke assumed that battle line commander Lee would welcome the opportunity. But Lee strongly opposed such an encounter. As it was he was not at all enthusiastic about such a night engagement with Japanese surface forces, believing that his crews were not adequately trained for it, whereas the IJN excelled at this sort of engagement. Spruance considered for an hour, then refused Mitscher's request to seek a night battle. Mitscher's staff was disappointed with Spruance's decision.[81]

As it was, the American fleet quickly devastated the Japanese airfields in the area. Few of Ozawa's pilots had a chance to fly to the land bases, however, as the majority were destroyed attempting to attack the American task forces.

American Admiral Raymond Spruance had arranged his warships in such a formation that permitted his fleet to have advanced radar warning of any incoming Japanese aircraft. Japanese pilots found it nearly impossible to evade the very effective radar-vectored combat air patrol. In the event, the U S. Navy flew more defensive sorties than the Japanese launched in total sorties of all kinds. The Japanese then were forced to fly directly through the immensely powerful anti-aircraft fire of the screening battleships and cruisers, which had been deployed across the front of the formation in a defensive line. American technological superiority, along with pilot skill, simply overwhelmed the Japanese. Especially skilled American fighter pilots exacted hideous losses from the Japanese attackers.

Ozawa launched morning search patrols at the first sign of daylight, using some of the 50 aircraft stationed on Guam. At 05:50 hours,

---

[80]Morison, Samuel Eliot, *New Guinea and the Marianas, March 1944–August 1944,* vol. 8 of *History of United States Naval Operations in World War II,* 1950. Admiral Morison's history is essentially the official U.S. Navy account of the war in the Pacific.

[81]Buell, Thomas B. *The Quiet Warrior: A Biography of Admiral Raymond A. Spruance.* 1987.

one of these aircraft located the American fleet. After radioing his sighting of U.S. ships, the bomb-carrying Zero attacked the destroyer Stockholm. but was shot down by destroyer *Yarnall.* Ozawa began launching the Guam-based aircraft for an attack. American radar on U.S. ships spotted the incoming Japanese aircraft when it was still a considerable distance away. A group of thirty Grumman F6F Hellcats from the carrier *U.S.S. Belleau Wood* defended. The Hellcats arrived while Japanese aircraft were still launching from Orote Field. Minutes later, additional radar contacts appeared on radar screens. These were additional forces being sent north from other Japanese-occupied islands. In one engagement a flight of 6F6 Hellcat fighters shot down 35 Japanese aircraft without the loss of a single Hellcat. This pattern would be repeated throughout the day. At 10:23, Admiral Mitscher ordered his carriers to turn into the wind on course east-southeast, and ordered all fighter aircraft aloft to await the Japanese. He then sent his bomber aircraft aloft to orbit open waters to the east rather than leaving them in a hangar deck full of aircraft vulnerable to a Japanese bomb attack.

The U. S. fleet picked up radar contacts 150 miles to the west at about 10:00. This was the first of the raids from the Japanese carrier forces, which was mounted with 68 aircraft. American carriers started launching every available fighter, and by the time they were in the air, the Japanese had closed to within 70 miles. Now the Japanese made their first major error in the battle. The Japanese aircraft began circling in order to reconstitute their formations for the attack. This ten-minute delay proved critical, and the first group of Hellcats met the raid, some 70 miles out from the first American picket ships. These 6F6 Hellcats were quickly joined by additional groups. Within minutes 25 Japanese planes had been shot down, against the loss of only one U.S. aircraft.

The next group of American fighters met those Japanese planes that had survived the first onslaught, and 16 more were shot down. Of the few remaining Japanese attack aircraft, some made attacks on the picket destroyers *U.S.S. Yarnall* and *U.S.S. Stockholm* but caused no damage. Three or four bombers broke through to Admiral Lee's battleship group, and one made a direct hit on the *U.S.S. South Dakota*, which caused many casualties, but failed to disable her. Not one of Ozawa's first wave broke through to the American carriers.[82]

At 11:07, American radar detected a second, much larger attack group of aircraft, consisting of 109 airplanes. The American fighters were met them while they still 60 miles out, and shot down at least 70. Six Japanese airplanes broke through and attacked Rear Admiral Montgomery's group, missing two of his carriers, but still causing casualties on each. Anti-aircraft guns destroyed four of the six. Another small group of torpedo planes attacked *U.S.S. Enterprise.* There were no direct hits, but one torpedo exploded in the wake of the ship. Three other torpedo-planes attacked the light carrier *U.S.S. Princeton* but were shot down before causing any damage. In all, U. S. defensive forces obliterated 97 of the 107 attacking aircraft.

Ozawa launched a third raid, consisting of 47 aircraft, which attacked from the north. Some 40 U. S. fighters met the Japanese at 13:00, while they were still 50 miles out from the task force. The U.S. fighters destroyed even Japanese planes, but a few broke through and made an ineffective attack on the *Enterprise* group. The pilots who flew this raid did not press home their attacks and thus suffered less than the others. About 40 of these aircraft returned to their carriers. *U.S.S. Bunker Hill* was nearly hit by a Japanese bomb during this attack.

The Japanese commander launched yet a fourth raid was between 11:00 and 11:30, but pilots had been given an incorrect position for the U. S. fleet and therefore failed to locate and attack it. They finally gave up, broke into two groups, and turned for Guam and Rota to refuel. However, the one group flying towards Rota stumbled upon Montgomery's task group. Eighteen aircraft joined battle with American fighters. The U. S. pilots shot down half of their number. A smaller group of nine Japanese dive bombers evaded U. S. planes and made attacks on the *U.S.S. Wasp* and the *U.S.S. Bunker Hill,* but failed to make any hits. American defensive response cut down eight of these aircraft. The larger group of Japanese planes that had flown toward Guam were most unlucky. Just as they were landing at Orote Field some 27 Hellcats appeared and destroyed most as they were landing.[83]

---

[82]Dickson, W. D., *Battle of the Philippine Sea.* London: Ian Allen, 1975.
[83]Smith, Douglas V. *Carrier Battles: Command Decision in Harm's Way.* U. S. Naval

Simultaneously, Ozawa experienced difficulties with American submarines. An American submarine placed a single torpedo in the aircraft carrier *Taiho.* Although that ship took only the single hit, poor damage control technique led to a catastrophic explosion which destroyed her. Opinion was that gasoline fumes caught fire and caused the detonation. That same afternoon, another American submarine torpedoed and sank the veteran aircraft carrier *Shokaku.* Follow-up strikes by American aircraft add the aircraft carrier *History* to the list of victims. Several other major warships damaged.

Admiral Ozawa was absolutely determined to force a showdown with the U.S. invaders. He had sent up 430 planes in four waves against ships under the command of Admiral Raymond Spruance. The result for the Japanese had been a disaster: in the first day of the battle the Japanese lost more than 200 planes and two regular carriers; and, as their fleet retired northward toward safe harbor at Okinawa, it lost another carrier and nearly 100 more planes. Some ninety percent of Ozawa's aircraft were rendered useless. The Japanese admiral had precious little to show for the sacrifice of so many pilots.

The American fleet sailed west during the night in order to be prepared to attack the Japanese at dawn. Spruance sent search patrols at first light. It was not until this morning that Admiral Ozawa learned of the disastrous results of the day before and that he still had only about 150 aircraft left. Nevertheless, he decided to continue the attacks. Lacking proper information, Ozawa thought that there were still hundreds of planes on Guam and Rota. Thus, he started planning new raids to be launched on June 21, largely with the use of aircraft that did not actually exist.

American search planes failed to locate the Japanese fleet until 15:40. That report made was so garbled that Admiral Mitscher knew neither what had been sighted, nor where. Then at 16:05, he received a clear report and Mitscher decided to launch a major strike even though there were only 75 minutes of daylight remaining. American aviators did not normally land at night because of the risk of significant losses due to landing mishaps.

Ozawa was able to put up only a handful of fighters to intercept

---

Institute Press, 2006.

the incoming U.S. attack. Later an estimate concluded Ozawa had no more than 35 defensive fighters. Fortunately for the Japanese, these few were skillful pilots, reminiscent of the earlier stages of the Pacific War. Too, the Japanese antiaircraft fire was intense. Still, the U. S. attack mounted some 550 planes, and most were able to press the attack. The first ships sighted were oilers, and two of these were damaged so severely that they were later scuttled. The American pilots then honed in on the carrier *Hiyō.* Four Grumman Avengers from *U.S.S. Belleau Wood.* hit it with bombs and aerial torpedoes and the *Hiyo* was set afire. The initial explosions were followed by a tremendous blast caused by leaking aviation fuel. Dead in the water, this ship slipped stern first under the waves, taking the lives of 250 officers and men. The rest of her crew of about one thousand, were rescued by Japanese destroyers.[84] The U.S. carrier aircraft attacked the carriers *Zuikaku, Junyō,* and *Chiyoda.* All were damaged by bombs, as was the battleship *Haruna.* Twenty American aircraft were lost in this strike.[85]

At 20:45, the first U.S. planes began to return to the American fleet. Knowing his aviators would have difficulty finding and landing on their carriers, Admiral Mitscher bravely decided to fully illuminate his carriers. He ordered the carriers to shine their powerful searchlights directly up into the night sky, despite the obvious risk of attack from submarines and night-flying aircraft. Picket destroyers joined in, firing star-shells to help the planes find the task groups. Despite this, eighty of the returning aircraft were lost. Some aircraft crashed on landing on the flight decks, but the majority of the crashes took place in the sea. However, many of the crews were rescued over the next few days by aircraft and submarines.[86]

Having already achieved a great victory, Spruance decided late on the second day not to press his attack further, a controversial decision to this day. During the two days of battle, U.S. losses totaled 130 aircraft and some damage to ships. There have been many theses written about

---

[84]Bryan, Joseph, and Philip Reed. *Mission Beyond Darkness.* Duell, Sloan and Pearce, 1945.

[85]D'Albas, Andrieu. *Death of a Navy: Japanese Naval Action in World War II.* Devin-Adair, 1965.

[86]Y'Blood, William T. *Red Sun Setting: The Battle of the Philippine Sea.* Annapolis, Maryland: US Naval Institute Press, 1981.

Spruance's refusal to follow through and chase down the remaining Japanese carriers. It is clear that Spruance showed a remarkable restraint, the exact opposite of Admiral Bill Halsey at the Battle of Leyte Gulf, as we shall see.

Much speculation has been made concerning the miserable showing by the Japanese aircraft at the Battle of the Philippine Sea. We need go no farther than looking at two factors. The first consideration may be singled out for special mention: the Japanese aircraft were extremely dated. They were among the very best at the beginning of the Pacific War, but Japan had stood pat. Meanwhile, other nations, notably the United States, had developed superior aircraft. Indeed, some designs, such as the 6F6 were designed specifically to meet the best Japan had. Japanese planes were highly maneuverable and had a longer range than U.S. planes, but they were inferior in several respects, particularly in their inadequate armor protection and lack of self-sealing fuel tanks.

Second, the pilots were inferior to that generation that had attacked Pearl Harbor and fought at Midway. Some Japanese pilots went into action with as little as three months of training. Initially, Japan had made the conscious decision to train a small group of highly talented pilots, rather than continue to prepare the next generations of fliers. When the first class was decimated, there were few pilots ready to take their place. By mid-1944, many U.S. pilots had spent two full years in training. The situation was turned about completely from December 1941 when it was the Japanese who filled the skies with highly trained pilots flying state-of-art aircraft.[87]

U.S. submarines also played an important but less publicized role in providing U.S. commanders with intelligence of enemy movements and in sinking Japanese ships. Japan never quite developed submarine warfare, undecided as they were what role these undersea craft should play in the Pacific War.

Many see the failure of IJN at Philippine Sea being related to the misuse of the air crews that Japan had to completely rebuild after the debacle at Midway. The Japanese carrier fliers were slaughtered almost to

---

[87]Tillman, Barrett, *Clash of the Carriers: The True Story of the Marianas Turkey Shoot of World War II.* Penguin, 2006.

a man. As a major fighting tool, the Imperial Japanese carriers had virtually ceased to exist. These once incredibly powerful implement of war had been utterly destroyed. This was the largest carrier-to-carrier battle in history.[88]

[88]Dull, Paul S. *A Battle History of the Imperial Japanese Navy, 1941–1945*. Naval Institute Press, 1978.

Battle of Leyte Gulf

The battle of Leyte Gulf, which took place from 22 to 26 October 1944, was the largest and most complex naval battle in history. It ended with a massive American victory that effectively destroyed the remaining fighting capability of the Japanese navy.

During the summer of 1944, General Douglas MacArthur prepared to fulfill his promise to return to the Philippine Islands. The original plan had called for a landing on Mindanao, the southernmost of the main Philippine islands. That would be followed by a larger invasion of Leyte, nearer the center of the islands. Finally, there would be the invasion of the largest of the islands, Luzon. In mid-June the Joint Chiefs of Staff recommended abandoning the invasion of Luzon and instead moving from Leyte to the island of Formosa, now called Taiwan. That was not what General MacArthur was willing to do. After all he had promised to return to the Philippines after being forced to leave the islands in 1942 just before capitulation. In late July General MacArthur, Admiral Chester Nimitz and President Franklin Roosevelt met on Hawaii. The result was that the original plan for the invasion of Luzon was confirmed. The campaign in the Philippines was to begin in December 1944.

Preliminary operations began on 6 September when aircraft from Admiral Mitscher's Task Force 38 bombarded the Palau Islands, 550 miles to the east of Mindanao. A small force attacked that island on 9-10 September, triggering a false alarm for the Japanese Navy. Next General MacArthur ordered an attack on the Visayan Islands, in the center of the Philippines, again throwing the Japanese into confusion. These adventures came off so well, and the Japanese were so confused, that Admiral William Halsey, commander of the U. S. Third Fleet,[89] suggested a new plan. Instead of waiting until December and then carefully advancing from

---

[89]Potter, E. B. *Bull Halsey*. Naval Institute Press, 2003.

south to north through the Philippines, he believed that an immediate invasion of Leyte would be successful. U. S. headquarters relayed this suggestion reached President Roosevelt who was then meeting at Quebec, Canada, with British Prime Minister Winston Churchill. These two leaders expressed approval of this idea. However, General MacArthur could not be contacted, because he had decided to accompany a force that was about to attack Morotai and that was operating radio silence. General MacArthur's chief of staff made the decision for him, and approved the new plan. Leyte would be invaded on 20 October.

General MacArthur's troops were to land on the good beaches on the east coast of Leyte. The engineers would immediately construct airfields inland which would be used to support both the fighting on Leyte itself and the landings on Luzon. The U. S. 7th Fleet would provide direct support, using both the big guns of the battleships and cruisers and also the aircraft from the several escort carriers. The 3rd Fleet would provide cover against any attempt by the Japanese navy to intervene.[90]

For their part the Japanese had spent the summer of 1944 working on a grand plan, another Great All-out Sea Battle. Operation Victory (Sho-Go) was a partially defensive, partly offensive, plan. The Japanese high command decided that at least one of four different American actions were likely. The most probable was an invasion of the Philippines. Still likely, but less probable was an invasion of Formosa to the south. The Japanese also considered the possibility of an invasion of the Kurile Islands in the north. Lest probable at this stage, although almost certain in the long run was a direct attack on the Japanese Home Islands.[91]

In the Japanese planning, Sho-1 was the defense of the Philippines and Ho-2 the defense of Formosa. The Japanese planners remained obsessed with the idea of the great decisive battle, which, if it resulted in a Japanese victory, could save national honor and prevent the ultimate invasion of the homeland. Victory in this grand battle was now Japan's

---

[90]Friedman, Kenneth. *Afternoon of the Rising Sun: The Battle of Leyte Gulf.* Presidio Press, 2001. See also Hoyt, Edwin P. *Leyte Gulf: The Death of the Princeton.* New York: Avon Books, 1987.

[91]Field, James A. *The Japanese at Leyte Gulf: The Sho operation.* Princeton University Press, 1947.

sole military hope.[92]

In the autumn of 1944 what remained of the once-great Japanese fleet was widely scattered. The main battle fleet was based at Gangling, a small island east of Sumatra and south of Singapore. This location had been chosen because it moored the fleet close to its main sources of fuel. So successful had been the Allied campaign against Japanese commercial shipping that there were few other options. The decimated carrier force had retreated to the Inland Sea in the Japanese Home Islands where a new generation of naval aviators was being trained since nearly all the previous generation had been killed in the Battle of the Philippine Sea.

The Sho-1 plan took advantage of this dispersal. Admiral Ozawa with his force of carriers was to approach the Philippines from Japan. His role as a decoy force was to draw the main American carriers and fast battleships away from Leyte Gulf leaving the invasion fleet vulnerable to attack. Admiral Ozawa clearly knew that the northern fleet was being deliberately sacrificed as an integral part of an overall strategy to win the great decisive battle. Admiral Ozawa still had a few carriers but he did not yet have a full complement of trained naval aviators. Neither was there time to train sufficient aviators, and there was a serious question whether IJN possessed enough instructors and sufficient fuel to do so.

Initially, Admiral Ozawa's role had been to take part in the main battle. But after the destruction of his last effective air groups in what the United States called the great Marianas Turkey Shoot, its role was downgraded to acting as the diversionary attraction.[93]

The main fleet at Lingga was to split into two parts. Admiral Kurita was to take the largest and most powerful part of the fleet, known as I Striking Force, through the middle of the Philippines. He was to emerge from the San Bernadino Strait, north of Leyte, and sweep south to attack the American invasion fleet. Admiral Kurita objected to the plan, not because he saw any flaws in the operation itself, but because he did not see the necessity of risking so much of his fleet for an attack on transport ships that would probably already have been unloaded by the

---

[92]Stewart, Adrian. *The Battle of Leyte Gulf.* Hale, 1979.

[93]Hoyt, Edwin P. *To the Marianas: War in the Central Pacific, 1944.* New York: Avon Books, 1983.

time the Japanese fleet arrived. The high command overruled his objections and he was told that he was to do his duty.

The second part of the main fleet, under Admiral Nishimura, was to pass through the Philippines further south and emerge from the Surigao Strait, between Mindanao and Leyte. He, too, was to attack the invasion fleet, in his case from the south.[94]

Last, Admiral Shima, with the smallest battle group, known as II Striking Force, was to sail from Japan via Formosa and join Admiral Nishimura. Initially, Admiral Shima had not been included in the Sho-Go plan but he had persuaded his superiors to allow his small force take part. The Japanese hoped that at least parts of the three southern fleets could break into Leyte Gulf and inflict crushing losses on the landing craft, supply vessels and smaller warships supporting the invasion of Leyte.

Admiral Halsey's 3rd Fleet was the main American striking force in the Pacific. At Leyte Gulf it consisted of fifteen fleet carriers, seven modern fast battleships, twenty-one cruisers and fifty eight destroyers. His larger ships were well armed for combat with Japanese battleships and carriers. American command had given Admiral Halsey contradictory orders. On the one hand it appeared that his main role was to find and destroy the Japanese fleet. Equally important was his job he to protect the vulnerable supply ships of the invasion forces moored in Leyte Gulf.

Admiral Kinkaid's 7th Fleet was dedicated to supporting the ground troops. He had sixteen escort carriers, six older battleships including several that had been sunk at Pearl Harbor and subsequently resurrected. He also had eleven cruisers and eighty-six destroyers. This was a powerful force, but did have one weakness. His carriers and battleships were armed for coastal bombardment, with high explosive shells and bombs, and carried very few armor piercing shells or anti-ship bombs. When Admiral Kinkaid found himself facing Japanese battleships his ships lacked the necessary tools.

Admiral Ozawa's Northern or Main Force was steaming from the Inland Sea in Japan, where his carrier air groups had been partially, but far less skillfully, reconstructed. He had four carriers, including the *Zuikaku,*

[94]Cutler, Thomas J. *The Battle of Leyte Gulf: 23-26 October 1944.* Harper Colins, 1994.

one of the best Japanese carriers of the war and a veteran of Pearl Harbor. The other three were all light carriers which had been constructed by converting support ships. These were the *Zuiho, Chitose* and *Chiyoda.* However, many of the aircraft allocated to this carrier force had been lost in the battle off Formosa on 12 through 16 October 1944. Thus, at Leyte Gulf he only had approximately 100 operational aircraft and only a few experienced air crews.

Admiral Kurita commanded the I Striking Force, which approached the Philippines from Brunei. At the heart of I Striking Force were the super-battleships *Yamato* and *Musashi.* These craft were, and remain today, the largest battleships ever constructed anywhere in the world. They were mounted with massive with nine fearsome 18.1inch diameter main guns. Admiral Kurita also had *Kongo* and *Haruna*, two pre-First World War battle-cruisers that had been turned into battleships in the late 1920s; and the *Nagato*, a battleship launched in 1919, but which mounted 16 in guns. This powerful force was supported by twelve cruisers and fifteen destroyers.

Admiral Nishimura, coming from Brunei, commanded two old battleships *Fuso* and *Yamashiro*, one cruiser and four destroyers. His chance of success could come only if Kurita or Ozawa had drawn almost the entire U. S. fleet north away from the Surigao Strait. Admiral Shima's II Striking Force, steaming from Formosa, was the weakest of the Japanese fleets, containing no battleships, three cruisers and four destroyers. This gave the Japanese a total of four carriers, nine battleships, nineteen cruisers and thirty-one destroyers.

The Japanese plan met with more success than a neutral observer might have anticipated. The fighting began when two American submarines discovered Admiral Kurita's force on 23 October 1944. It was known as the Battle of the Sibuyan Sea, and it was fought on 23 and 24 October 1944. On 23 October the submarines sank two Japanese cruisers.

The battle of Leyte Gulf began well to the west of the Philippines on 23 October (battle of the Sibuyan Sea, 23 to 24 October, although this battle began outside that sea). Darter and Dace, two American submarines, found Kurita's fleet while it was sailing along the north coast of Palawan Island in the South China Sea. The American subs attacked Kurita and sank two cruisers, including his flagship *Atago*. A third cruiser was crippled and had to return to Brunei, taking two destroyers as a screen.[95]

On 24 October, American aircraft, alerted to the presence of this fleet, sank the battleship Musashi. Admiral Halsey's scouts then detected Admiral Ozawa's carrier force and decided to head north to deal with this apparent threat. Halsey swallowed the bait, taking hook, line, and sinker. On the morning of 24 October, a Japanese scout plane based on Luzon found Task Force 38, sailing to the east of the island. The Japanese Navy had more aircraft based on Luzon than on Ozawa's carriers, and during the morning of 24 October just over 200 Japanese land based naval aircraft attacked the task force. For about an hour the Japanese were fought off, but just as the main attack ended the light carrier Princeton, part of TG 38.3, was hit by a single Japanese dive bomber. Prolonged efforts to save the carrier failed and eventually she was sunk by American torpedoes. Most of her crew survived, but an explosion caused heavy casualties on the cruiser Birmingham, one of the ships taking part in the fire-fighting effort.

Next Ozawa's carrier aircraft made an appearance. At about 11:45, about two thirds of his aircraft attacked Halsey's fleet but without any success. The inexperienced carrier aviators then flew on to land on Luzon. At this point Halsey probably didn't realize that these aircraft came from a carrier force, but Ozawa was finally located by American scout planes in the afternoon

The Japanese came closest to success in the center. With Halsey in the north chasing the decoy fleet and Kinkaid in the south the northern approaches to Leyte Gulf were only protected by escort carriers, destroyers and destroyer escorts. Kurita emerged from the San Bernardino Strait, turned south, and ran headlong into Taffy 3, one of three task groups of six escort carriers. The American carriers turned and tried to reach relative safety, launching their aircraft as they sped away. The destroyers and destroyer escorts made valiant attempts to halt the Japanese advance,

---

[95] Willmott, H. P. *The Battle Of Leyte Gulf: The Last Fleet Action.* Indiana University Press, 2005.

while the jeep carrier aircraft made repeated attacks on the Japanese battleships. Most of the aircraft carried ordnance suitable for support of the army and not anti-ship materials. One escort carrier was quickly sunk, but these tiny vessels accomplished their mission. Kurita emerged unnoticed from the San Bernardino Strait, turned south and headed towards Leyte Gulf. He then found Admiral Sprague's Taffy 3, of six escort carriers, three destroyers and four destroyer escorts. Sprague conducted a skillful fighting retreat, harassing the Japanese with his aircraft even though they had no armor piercing bombs and destroyers. During the fighting, the carrier *Gambier Bay* was sunk as were three of the escort vessels. Sprague's small group was close to defeat when Kurita decided to withdraw from the battle, reform his fleet and resume his advance into Leyte Gulf. The Japanese also suffered major losses including three cruisers that were sunk on the day along with a fourth cruiser that was badly damaged. [96]

Kurita turned away at 9: 11 am. It took two hours for his fleet to regroup. He then turned south and headed towards Leyte Gulf and the American shipping. At about 11: 40 am his scout made a false report of sighting a battleship and Kurita turned aside to try to engage it. Kurita spent the rest of the day chasing ghosts off Samar, before giving up and retiring through the San Bernardino Strait. He then turned south again, before at 12.35 he decided to turn back north and try and find an American carrier group believed to be 100 miles to his north. In fact Halsey's carriers were much further north, and out of range. Kurita steamed north all afternoon in an attempt to find this phantom force, before at around 6 pm he finally gave up and made his way back into the San Bernardino Strait heading west. The Japanese suffered major losses, including three cruisers that were sunk on the day along with a fourth cruiser that was badly damaged.[97]

Totally misinterpreting the size and nature of his valiant opponents, Kurita, who had lost some perspective on the battle when his

---

[96] See two books by Como, Byron. *The Defenders of Taffy 3*. Seattle: Amazon, 2017; and *The Raiders of Taffy 3*. Seattle: Amazon, 2017.

[97] Cox, Robert Jon. *The Battle Off Samar: Taffy III at Leyte Gulf*. Agogeebic Press; 5th ed., 2010.

flagship was damaged, then decided to regroup and return to his original task in Leyte Gulf. Thus, the Battle of Samar. which might have succeeded beyond IJN's wildest expectations, was lost.[98]

Once again, the IJN, while seeking that one great final all-out sea battle, snatched one final defeat from the jaws of victory. With just a bit of daring Kurita could have inflicted extremely serious defeat on the Leyte landing force and invasion fleet. Taffy Three had far more real damage on the powerful Japanese force than anyone could possibly have imagined. That tiny American group displayed all the nerve, courage, heroism and bravery that Kurita lacked.

Further south, General Tominaga's army air corps attacked the Seventh Fleet in Leyte Gulf, but with little effect, although the escort carriers were forced to concentrate on air defense instead of close support. The Japanese lost around 70 aircraft in this attack.

Halsey's carriers went on the offensive on 24 October, launching five separate air strikes against Kurita, flying sorties between 9 am and mid-afternoon. The principal object of these attacks, in what was known as the Battle of the Sibuyan Sea, was the super-battleship *Musashi.* Halsey's men hit it with multiple torpedoes and bombs. They also hit a heavy cruiser which was so badly damaged that it was forced to turn back. At around 15.30, Kurita decided to temporarily turn back to avoid coming under aerial attack in the narrow San Bernardino Strait. The American observers believed that Kurita might be retiring from the area. Instead after just under two hours he turned back east and headed into the San Bernardino Strait unobserved.

The key day of the Battle of Leyte Gulf was 25 October when three separate battles were fought. In the north Halsey sank all four Japanese carriers in an engagement known as the Battle of Cape Engano. The carriers were, in effect, empty shells without a reasonable complement of skilled aviators.

In the south, Admiral Kinkaid intercepted and destroyed Admiral Nishimura's fleet in the Battle of Surigao Strait, performing the classic maneuver known as crossing the T. Admiral Shima then decided to turn back. In the south the fighting was equally one sided. Admiral Oldendorf,

---

[98] Hornfischer, James D. *The Last Stand of the Tin Can Sailors.* Bantam, 2004.

with the six old battleships of the 7th Fleet, blocked the exit from the Surigao Strait. Nishimura's ships were attacked by PT boats in the strait and by torpedoes from American destroyers as they approached the exit. One battleship and two destroyers were sunk and a third forced to turn back. By the time Nisihimura reached the American battleships he only had one battleship, one cruiser and one destroyer. In the resulting gun battle the battleship was sunk and the cruiser very badly damaged. It escaped for the moment but was sunk while attempting to escape. Only the destroyer reached safety. Shima realized the battle was lost and turned back, saving his ships. Their escape was aided by news from the north, where Kurita's battleships had emerged into Leyte Gulf. Oldendorf had to cancel the pursuit and turned north to prepare for a possible second battle.[99]

Halsey now had a choice to make. His orders from Nimitz were to protect the invasion fleet unless a chance came up to destroy the Japanese fleet. He now knew of three Japanese forces. To the north was Ozawa with four carriers and two battleships. To the west was Kurita, who had been battered all day, lost the biggest battleship in the world and begun a possible retreat. To the south was Nishimura with a force that wasn't really a threat to Kinkaid's 7th Fleet. Halsey decided that he could combine his two tasks by leading his 3rd Fleet north to destroy the Japanese carrier force. In every earlier battle of the Pacific War this would have been the correct decision - the carriers were now the most dangerous weapon in the naval armory and the highly skilled aviators of Pearl Harbor and the year that followed could have caused havoc if they reached the invasion fleets. Halsey's real mistake was that he failed to make sure that someone was watching Kurita and the San Bernardino Strait. He and Kinkaid both assumed that the other fleet was carrying out that task but neither did.

Halsey has also been criticized for deciding to attack the Japanese carrier fleet in the first place, largely on the grounds that it was carrying very few aircraft, although there was no way that Halsey could have known this. The Japanese had only recently deployed large numbers of naval aircraft in the battles off Formosa, and Halsey had just been attacked twice by naval aircraft on 25 October. As far as he knew, the four carriers

---

[99] Woodward, C. Vann. *The Battle for Leyte Gulf.* (1947) Skyhorse Reprint, 2007.

and two modified battleships in the north were all carrying their full complement of aircraft. Halsey's mistake was not making sure a suitable force was watching the San Bernardino Strait. There is little doubt had the Japanese broken through Taffy Three's defense that Halsey would have been court-martialed. A piece of U. S. Navy slang for Halsey's actions is Bull's Run, a phrase combining Halsey's newspaper nickname "Bull" even though in the U. S. Navy he was known as "Bill" Halsey, with an allusion to the Battle of Bull Run in the American Civil War.[100]

Naval historian Admiral Samuel E. Morison wrote that, if a portion of Halsey's fleet "had been detached a few hours earlier, after Kinkaid's first urgent request for help, and had left the destroyers behind, since their fueling caused a delay of over two and a half hours, a powerful battle line of six modern battleships under the command of Admiral Lee, the most experienced battle squadron commander in the Navy, would have arrived off the San Bernardino Strait in time to have clashed with Kurita's Center Force ... Apart from the accidents common in naval warfare, there is every reason to suppose that Lee would have "crossed the T" and completed the destruction of Center Force."

This was the last major naval engagement of the Pacific War. Japan had only a few ships remaining, although one was the second super-battleship. The Battle of Leyte Gulf had been Japan's final opportunity to fight the Great All-out Sea Battle. As with the earlier opportunities they blew it completely. One may seriously doubt that Alfred Mahan would have thought that any nation would have had three such opportunities.[101]

It can be said that the Battle of Leyte Gulf was decisive in that it destroyed nearly all the remaining functional Japanese surface fleet and therefore essentially ending Japan's ability to move resources from Southeast Asia to the home islands. Japanese losses included four aircraft carriers, three battleships, six heavy and four light cruisers, and eleven destroyers, along with several hundred aircraft and over 10,500 sailors. Allied losses were one light carrier, two escort carriers, two destroyers and one destroyer-escort. Despite overall failure, however, the Japanese

---

[100] Cutler, Thomas. *The Battle of Leyte Gulf: 23–26 October 1944*. Annapolis, Maryland, U.S.: Naval Institute Press, 1994.

[101] https://www.britannica.com/event/Battle-of-Leyte-Gulf

showed that with determination they could still press home attacks against an Allied armada with huge technical and material advantages.[102]

[102]https://www.history.com/topics/world-war-ii/battle-of-leyte-gulf. See also Hoyt, Edwin P. *The Battle of Leyte Gulf: The Death Knell of the Japanese Fleet.* New York: Pinnacle Books, 1972.

Japanese Small Arms

Japanese markings on firearms are typically a combination of the Emperor's mark, or Chrysanthemum, a marking denoting the arsenal at which the gun was manufactured, a serial number in Arabic numerals, and the model number in Japanese script. Japanese military equipment was researched and developed along two separate procurement processes, one for the Imperial Japanese Army and one for the Imperial Japanese Navy. Most of the Japanese small arms looked and functioned like substandard versions of British and German weapons. They were sufficient only against even more poorly equipped Chinese troops.

In infantry weapons, the Japanese lagged far behind Germany, Great Britain, and the United States. Only the Italian Army was equipped with weapons of similarly inferior design and construction. The quality of Japanese infantry weapons of World War II ranged from being good to mediocre to very bad. A very few of them were good, most were mediocre, and some were very bad. Japanese rifles were obsolete compared to rifles used by other major powers because their design dated back long before the Great War and remained relatively unchanged throughout the war. Japanese pistols ranged from being quite mediocre to being very bad. Compared to their European and American counterparts, they were inferior. The Japanese military developed a pistol that was aptly regarded as a piece of garbage: the Type 94 Nambu pistol. It was ugly, of poor quality, and had one peculiar feature that proved particularly dangerous for its operator: an exposed sear that could fire the gun accidentally.[103]

Japanese small arms, like many of the nation's armaments, were overly complicated, difficult to operate, and rarely upgraded. Those small arms that were once competitive began to age rapidly. Even the Chinese

---

[103] Drea, Edward J. *In the Service of the Emperor: Essays on the Imperial Japanese Army*. Bison Books, 2003. See also his *Japan's Imperial Army: Its Rise and Fall, 1853-1945*. University Press of Kansas, 2009.

troops armed with versions of the reliable and time-tried and true Mauser 98 rifles had better infantry rifles than did Japanese infantry. And the Japanese had absolutely nothing comparable to the U.S. M-1 Garand rifle. If the authors had had their choice of any standard infantry weapon of World War II it would have been the Browning designed BAR. A close second choice would have been the Johnson semi-automatic rifle, and again Japan had no equivalent arm.[104]

One of the odd failures of the Japanese military was the lack of waterproofing to cartridges issued for jungle warfare. Just as Japanese troops had essentially no jungle training, largely because there were no jungles in Japan, so did the Japanese ordnance misunderstand the needs of troops for waterproof ammunition.[105]

A recent expert has summed up the state of Japanese small arms in the years prior to, and including World War II,

> Much has been written that the Japanese infantry weapons of World War II were poorly designed and manufactured and ineffective in combat. During the 1930s, the Japanese high command falsely believed that an army based on the Bushido code would not be hampered by Japan's inadequate industrial base because it required neither state-of-the-art mechanization nor a cumbersome logistical tail. A reliance on material goods, necessitating an extensive supply network, was viewed by the dominating forces within the Japanese high command as a modern evil that could destroy the fighting spirit of the IJA. The IJA high command consistently resisted weapons modernization, fearing that it would lead to the infantry's abandonment of tradition of hand-to-hand combat to win the decisive victory. Thus, the general staff approved the design of the infantryman's weapons based on close-order combat, where he was programmed to always advance, keeping the enemy unnerved and off balance.[106]

---

[104] Tantum IV, W. H.; and E.J. Hoffschmidt, eds. *Second World War Combat Weapons*, Volume 2. *Japanese Tanks, Uniforms, Artillery, Small Arms, Medals, Grenades, Ordnance and Markings*. WE Inc., 1968.

[105] Isomura, Teruaki. *Early Made Japanese Military Small Arms Ammunition*. Published by Author, 1984.

[106] Diamond, Jon "Arisaka Rifles of the Imperial Japanese Army" Warefare History Network, https://warfarehistorynetwork.com/ 2018/12/18/arisaka-rifles-of-the-imperial-japanese-army/

Arisaka Rifles and Carbines.

The basic Japanese infantry rifle was known as the Arisaka, manufactured in both 6.5 mm and 7.7 mm cartridges. Arisaka was a family of Japanese military bolt-action rifles in production and use beginning in approximately 1897, when they replaced the Murata rifle. Arisaka rifles in both calibers were used until the end of the Pacific War in 1945. The Arisaka rifle was designed by Colonel Arisaka Nariakira (1852–1915), who was later promoted to lieutenant-general and also received the title of baron from Emperor Meiji, in 1907. Arisaka was born to a samurai retainer, but at the age of 11, he was adopted by firearms craftsman Arisaka Nagayoshi, from whom he took his family name. After the Meiji restoration, Arisaka enlisted in the Imperial Japanese Army. In 1891, he caught the attention of General Murata Tsuneyoshi, designer of the then standard Japanese Army rifle named the Murata after him. Subsequently Arisaka was appointed to a position in the Tokyo Arsenal.[107]

In 1897, Arisaka completed work on the Type 30, an improved version of the Murata Rifle, which was adopted by the Japanese Army as its standard weapon. It was first deployed during the Chinese Boxer Rebellion, 1899-1901. The Imperial Japanese Army regarded the Type 30 Rifle as underpowered. The Type 31 guns lacked recoil buffers and had poor accuracy. As a result of engaging the Chinese in 1894, IJN discovered that their current rifles were markedly inferior to their enemy's Mannlicher Gewehr 88. As a result, Colonel Arisaka went to work designing the Type 38 rifle to replace the outdated and expensive to produce Murata rifle. The Arisaka Type 38 rifle chambered in the 6.5mm cartridge (1905) was known to the Japanese soldier as the *sanpachiju.*

In 1903, Arisaka was appointed head of the Army Technical Bureau. He oversaw a committee charged with improving older models, including the Type 30 Rifle. The chief designer on the project was Captain Nambu Kijiro, who later designed the series of semi-automatic pisrols that bear his name. The result of this joint effort was the Type 38 rifle, generally

---

[107] *Journal of the Military Service Institution: Military Libraries - New Small Calibre Rifle - Military Service of Indians - The Mexican Army - West Indies and Nicaragua Canal - Chinese and Japanese Armies - War Between China and Japan.* Military Service Institute, 1894.

known as the Arisaka Rifle. The Army issued the Arisaka to its front-line infantry beginning near the end of the Russian-Japanese War of 1904-1905. It was a five-shot weapon that used an internal box magazine loaded with 6.5mm cartridges via brass or steel stripper clips. It had a bolt-action system patented by Mauser. It was a reliable weapon with a weight of nine pounds (empty), relatively light for its length of over four feet (50.25 inches), which was greater in length than either the future M-1 Garand or Model 1903 Springfield rifle used by American infantry. The Type 38 rifle had an unusually long barrel to gain acceptable accuracy, and at 31.4 inches it produced little recoil. Its production dated back the Russo-Japanese War in 1905, and it remained continuously manufactured until 1945, during which time over three million were made. Thus, like many other belligerents, the Japanese utilized rifles that were previously used during World War I. Combat troops reported favorably on the new rifle, especially its ruggedness. The Type 38 Rifle, and its various modified versions, continued to be used by the Japanese military until the end of World War II.[108]

The most common specimens and the Type 38 chambered for the 6.5 x 50 mm cartridge, and the Type 99 chambered for the 7.7 x 58 mm cartridge, which was comparably powerful to the U.S. 30/06 round. Only Italy and Japan used 6.5 mm [.264 caliber] cartridges while other nations used .30 caliber or slightly larger cartridges. Among the several serious problems encountered with the Arisaka series is the safety which is located at the end of the bolt. It is very difficult to operate and cannot be changed from safe to fire positions without removing the rifle from a shooting position. The Arisaka series of rifles was the best, and essentially only, infantry rifle that the Japanese had during the Pacific War, despite each model's differing calibers which can lead to logistical problems since all have been used at more or less at the same time as well as cumbersome loading procedure[109]. Japan also used 7.7 mm cartridges in rimmed, semi-rimmed, and rimless versions, none interchangeable with the other two.

[108]https:// warfarehistorynetwork.com / 2018 / 12 / 18/arisaka-rifles-of-the-imperial-japanese-army/. See also https:// www.militaryfactory.com / smallarms / detail.asp?smallarms_id=373

[109] Duncan O. McCollum, *Japanese Rifles of World War II*. Excalibur Publications, 1996; See also Fred L. Honeycutt, *Military Rifles of Japan*, Julin Books, 1996.

Only close examinations can differentiate.[110]

The 50-inch-long Arisaka Type 38 6.5mm (1905) rifle in actual use proved to be too long for the typically short Japanese infantryman. Many infantrymen had difficulty reaching the bolt when the butt was at the shoulder in a firing position. Firing the rifle proved to be difficult for the diminutive Japanese soldiers to aim and rapidly fire in the jungle. Its nine-pound weight, added to its length, made the weapon at best marginally suitable for jungle warfare. It was generally accurate so it was acceptable for close-quarter action in the jungle. The long, somewhat unstable, bullet tended to tumble when it hit its target, making wounds much worse than would have been the case if the bullet traveled in a straight path.[111]

The Arisaka Type 38 6.5 mm rifle was also made in a short version with the reduced overall length of 44.5 inches and weighing in at 8.5 pounds, only half a pound less than the longer version. Some of these Type 38 short arms were issued to infantry, but only late in the war. Although it was quite popular for jungle fighting because of its shorter overall length, most were issued to soldiers of supporting arms and logistic services. Type 38 carbine was simply a shortened type 38 rifle. A variant, for paratrooper use, was made with a hinged stock through the wrist, although it was not particularly successful.[112]

The Type 44 carbine was introduced in 1911. It had a permanently fixed folding bayonet. Caliber was 6.5. The Russo-Japanese War of 1905 proved that a practical cavalry carbine should be added to the INJ's arsenal of weapons. IJN designed such a short rifle with specifications that were essentially identical to the longer Arisaka Type 38 rifle n 6.5 x 50 mm caliber. The carbine was 38.25 inches long and weighed 8.8 pounds. As a stop-gap measure this arm worked far better than its predecessor. However, the IJN redesigned the cavalry carbine so it was better suited for the use of mounted troops. Thus, IJN deployed an Arisaka Type 44 (1911) cavalry bolt-action carbine that was actually the same as the earlier Type

---

[110] https://www.militaryfactory.com/smallarms/detail.asp?smallarms_id=249

[111] Allan, Francis C. and Macy, Harold W. *The Type 38 Arisaka.* AK Enterprises, 2007.

[112] Skennerton, Ian D. *Japanese Rifle 6.5mm Arisaka Type 38.* Arms & Militaria Press, 2008.

38 carbine model, except for having a folding bayonet that was permanently attached to the weapon.

It had the same overall length of just over 38 inches, and a weight of just over 8.8 pounds, but now the cavalryman would no longer have to ride with his bayonet secured to his belt. The biggest drawback was the excessive muzzle weight, making it difficult to aim, thereby diminishing the weapon's accuracy. Although designed for cavalry user, due to its more compact design, the Arisaka Type 44 (1911) cavalry bolt-action carbine was the weapon of choice for troops destined for the jungle. Since cavalry units were obsolete by the time the Pacific War opened there was no reason not to issue the Type 44 to infantry. It would have fit well into jungle warfare, a place where long-range shooting was all but unnecessary and its shorter length made it easier to handle. Reportedly, high manufacturing costs terminated the production of this rifle in 1942.[113]

Type 97 rifle was made specifically for sniping. The Imperial Japanese Army adopted a standardized sniper rifle in the form of the Type 38 bolt-action infantry service rifle which, when a 2.5x scope and monopod was added and the stock reduced in weight, became the "Type 97." The series was introduced during 1937 in time for the bulk of the fighting in the Pacific War. The Type 97 differed little from its Type 38 origins. One of the key limiting factors was its continued reliance on the 6.5x50 mm Arisaka rifle cartridge which fared somewhat poorly when compared to its European counterparts. The rifle was conventional in that a manually-actuated bolt-action system was used and feeding was from a 5-round internal magazine (stocked by way of "stripper" clips). Overall weight was 8.7 lb. with an overall length of 50 "and barrel assembly measuring 31.4" long. Production of the Type 97 spanned from 1937 until the end of World War II. In practice, the rifles proved their worth. Even though the 6.5mm Arisaka rifle cartridge was relatively underpowered, it was lethal nonetheless and the added inherent benefit was that little muzzle flash was encountered, making the sniper's position hard to identify particularly in deep brush or in the jungle environment. The Type 2 [Type 99 sniper rifle] was chambered for the 7.7 cartridge and was mounted with

[113]https://www.militaryfactory.com/smallarms/detail.asp?smallarms_id=547

either 2.5 or 4 power telescopic sight.[114]

The rare Type MO rifle was supplied for the 8x57 Mauser cartridge. It was made in Germany and imported in very limited numbers.[115]

The Type 2 Terra (Paratroop) rifle was a take-down rifle made for paratroopers. The Type 2 Paratroop rifle is basically a take-down version of the Type 99 Arisaka rifle, loading the same 7.7mm X 58 Arisaka cartridge. The Japanese experimented with several paratroop rifles, including ones with folding stocks and an interrupted thread-style take-down before settling on the Type 99, which has a screw-in wedge that holds the two halves together. The Type 2 is the only paratroop rifle that was produced in any real quantity, and even it is rare, with only about 21,200 produced. The model designation comes from the year based on the Japanese system that counted from the date of the mythological founding of the Japanese Empire by Emperor Jimmu in 660 B.C. By this system, the Western year 1942 when the gun was adopted was 2602, with the "two" becoming the model designation. Actual production began in late 1943.[116] The Japanese did not make very extensive use of paratroopers so the development of specialized weapons was ignored. However, the experiences of the airborne attack on Palembang in Sumatra, Dutch East Indies, on February 13-15, 1942 may have provided the impetus for the Imperial Japanese Army to create paratroop weapons. In that battle, the airborne troops were equipped with standard rifles that were dropped in separate canisters and ended up landing in swamps some distance from the men, who then had to fight with just pistols, bayonets and grenades.[117] It

---

[114] Bruderlin, Edward B & Robert S. Nelson. *Research, Development and Production of Small Arms and Aircraft Armament of the Japanese Army*. Ordnance Technical Intelligence Report Number 19. Published by General HQ, Army Forces., Pacific, 1971.

[115] Honeycutt Jr., Fred L. and Anthony, F. Patt. *Military Rifles of Japan.* Fifth edition; Julin Books, 2006.

[116]http://www.nambuworld.com/type2paratrooppix.htm. See also *Japanese Parachute Troops.* Wickenburg, AZ: Normount Technical Publications, Wickenburg, AZ, pp. 46-47. This is a reprint of a US Army manual under the "Combat Bookshelf" imprint.

[117] Honeycutt, Fred and F. P. Anthony. *Military Rifles of Japan*, Julin, 1996, pp. 134-135.

was created because the earlier Type 0 and Type 1 Terra rifles made for paratroopers; rifles simply did not work well. The Type 100 Terra Rifle had a detachable barrel for paratrooper use. 7.7 caliber

The Type 99 rifle was chambered in 7.7 Arisaka and adopted in 1939. No other rifle, civilian or military, was fitted with so many gadgets. Among the unusual and basically useless add-ons was the sight designed to assist the shooter to lead enemy aircraft. The mono-pod arguably might assist in steadying this overly long rifle when the soldier was firing from the prone position. Otherwise it added a small amount weight and caught on brush and even grass. The highly varnished finish might help protect against jungle rot, but it also reflected light helping the enemy to locate its bearer. The dust cover, also featured on some earlier rifles, rattled and occasionally gathered debris making loading and ejecting cartridges difficult.[118]

Type 99 carbine was a shortened version of the rifle. During the Korean War, approximately 126,500 short and 6,650 long Type 99 Rifles were re-chambered under American supervision at the Tokyo arsenal to fire the U.S. Military M-2 caliber .30-06 Springfield cartridge. These rifles were fitted with a lengthened magazine well and had a small notch cut in the top of the receiver ring in order to accommodate the slightly longer length of the 30-06 round. The bayonet is covered briefly, as is the different types of .30-06 ammunition that can be fired out of the rifle (M-2 Ball, Tracer, Armor Piercing, blank). The section on the antiaircraft sights explains how the sights are to be used and which marks on the sight arms correspond to speed and distance. In other words, the narrative explains how to use the sights to shoot at aircraft.[119]

Type 2. This shoulder arm was chambered in 7.7mm. It was fixed with an interrupted screw breech method of take down, supposedly for

---

[118] *Japanese Rifles of World War II*. This pamphlet type publication covers the Model 38, Model 99, and Model 30. It is an informative examination of the various rifles employed by the Japanese Imperial Army during World War II. Rifles discussed in this manual are the 6.5mm Model 38, the 7.7mm Model 99 the Sniper's 6.5mm Rifle & The Long 7.7mm Rifle. Includes: general info; id. & characteristics; operating; ammo; maintenance; & accessories. Illus, 34 pgs. No publication information was included.

[119] Skennerton, Ian D. *Japanese 7.7mm Arisaka Type 99 Rifle*. Philadelphia: Ray Riling Arms Books, 2008.

paratrooper use.[120]

Type One modified Italian Carcano was the most unusual arm issued in Japan during the Pacific War. The Imperial Japanese Army ran the Japanese arsenals and was allocating all of their production capacity for its own use. When the war in China heated up in 1937, the Imperial Japanese Navy could not get enough domestically-produced rifles from the Army-controlled arsenals, so they went to Europe and looked at foreign sources. The IJN liked German rifles but found them too expensive so they signed contracts with the Czechs and Italians. The Italian order was for 60,000 rifles, half made by RSFAE, one-quarter by Brescia and one-quarter by Beretta. The rifle has a Carcano action but is otherwise configured to look and handle like a Type 38. The caliber was the standard 6.5mm X 50 Arisaka. These arms seem to have seen relatively little use and many were found warehoused at the end of the war. The Italian Carcano was the worst infantry weapon of the Second World War with precious little to recommend it irrespective of caliber.[121]

Sniper rifles are rather standard among modern armies. Long range shooting with great accuracy against specific targets as a feature of all modern armies. For that purpose, all militaries have developed specialized rifles, usually mounted with magnifying optical sights. For its sniping, IJN issued a 2.5x Tokia telescopic scope which was mounted on the left side of the receiver behind the magazine breach on the Type 38 rifle. Developed in 1937, this was referred to as the Type 97 sniper rifle and used a smaller 6.5mm cartridge. However, the performance of this gun for long-range marksmanship left a lot to be desired. It was the result of a development program that extended over 10 years and essentially produced only an Arisaka Type 38 rifle with an added telescopic sight. The sight was mounted so low above the action that the bolt lever had to be lengthened and angled downward, while the sight was offset to the left so that the shooter could still operate the bolt and use the ammunition charger.

---

[120]Arisaka, Nariakira. *Technical Manual for Rifle Type 99 Japanese*. Reprint by Middle Coast Publishing, 2017.

[121]Smith, W. H. B. *Small Arms of the World: A Basic Manual of Small Arms*. Harrisburg: Stackpole Books, 1969. See also Weeks, John. *World War II Small Arms.* New York: Galahad, 1979. See photos at http://www.nambuworld.com/typeipix.htm.

With the Type 97's reduced performance as a marksman's weapon, the Japanese infantry-sniper doctrine adapted to the weapon's deficiencies and focused on its snipers perfecting camouflage and concealment. The Type 97 sniper rifle's low muzzle flash and smokeless propellant were effective in medium-range sniper action where firing positions would be less conspicuous. A sniper version of the Arisaka Type 99 7.7mm rifle was issued in 1942 and was fitted with either a 2.5x or 4x Tokia telescope, but this gun did not get its own designation.[122]

Type 5 Semi-automatic Rifle

Japan produced an experimental copy of the Garand, which they called the Type 5. The Type 4 / Type 5 rifle was a Japanese copy of the classic American M1 Garand semi-automatic service rifle. Throughout the war, the Japanese infantryman had relied on proven, but decidedly limited, bolt-action types. To compete with American troops armed with the M-1 Garand required a new, more modern weapon with a smooth repeat-fire action. By the mid-1930s, Americans had both moved on to repeat-fire rifles and the Germans and Soviets were following suit with designs all their own. Rather than invest in an all-new weapon, Japan selected the proven Garand type. While the Japanese had toyed with several semi-automatic infantry rifle designs, none of these projects supplied a viable end product. The commitment to the Pacific War did not allow for a major change in basic armament.

The gun is recognized under the designations of "Type 4" and "Type 5" in sources and is far better known as the Japanese Garand. This small arm stole the key elements making up the Garand rifle, including its form and function. The Type 4/5 was chambered for the standard Japanese 7.7x58mm Arisaka rifle cartridge. Unlike the Garand the Japanese 5 semi-automatic service rifle made use of a ten-round internal magazine as opposed to an eight-shot design which required an en blanc magazine to function, a decided drawback. The longer Japanese magazine case

---

[122]Bruderlin, Edward B & Robert S. Nelson. *Research, Development and Production of Small Arms and Aircraft Armament of the Japanese Army*. Ordnance Technical Intelligence Report Number 19. Published by General HQ, Army Forces., Pacific, 1971.

projected from the bottom of the gun body. One loaded the magazine with five-round stripper clips. The Type 4/5 rifle utilized a leaf-style rear sighting device as opposed to the Garand's aperture arrangement. On the whole, the Japanese approach remained largely faithful to the American design. With design work beginning in 1944, Japanese factories began setting up for mass production of the "Japanese Garand" in 1945. However, the Japanese surrender of August ended all planned manufacture of this semi-automatic rifle and only about 100 examples were completed from the 250 that were in the works at war's end.

Twenty specimens were known to be confiscated by the Allies at the Japanese Naval Arsenal on Honshu after the cessation of hostilities with one displayed at the Cody Museum. Obviously, the rifle came far too late to any bearing on the war effort.[123]

Handguns

Weapons produced by Nambu were considered the worst since they were prone to misfire and were quite heavy for their weapons class. The Nambu pistol was designed by Lieutenant General Kijiro Nambu. Nambu claimed the design originated with experimentation during the 30-year Automatic Pistol Plan of 1897 in Japan. The pistol series was influenced by the German Mauser C-96. A Japanese commission had toured Europe and reported recent developments in small arms designs. The first Nambu type known as the Type A was completed in 1902. The original Nambu was the Type A, designed by General Kijiro Nambu in 1902. The Type A had two basic variants with the first group of Nambus produced between 1903-06 being quite different from those produced after 1906. Type A underwent trials by the Imperial Japanese Army but was never formally adopted. Many Original Type As were then sold commercially to China and Siam. The Nambu Type A Modified pistol was adopted by the Imperial Japanese Navy in 1909 and by the Thai Army in the 1920s. Type B Nambu was a three-quarters sized scaled-down version of the Nambu Type A Modified. The Nambu Type B was produced because

[123] Tantum IV, W. H.; and E.J. Hoffschmidt, eds. *Second World War Combat Weapons*, Volume 2. *Japanese Tanks, Uniforms, Artillery, Small Arms, Medals, Grenades, Ordnance and Markings*. WE Inc., 1968.

the Type A Nambus were felt to be too cumbersome and sold poorly. Japanese army officers were expected to purchase their own side arms. Type B sold for ¥180 when a first lieutenant's monthly salary was ¥70. The Type B Nambu was commonly referred to as the Baby Nambu in Western Markets because of its small size. It fired a 7mm round than the conventional 8×22 mm Nambu. Type 14 Nambu was designed in 1925 with steps taken to simplify manufacture, thereby reducing cost. It was officially adopted for issue to non-commissioned officers of the Imperial Japanese Army in 1927. Army supply also made them available for purchase by officers. Type 14 was an improved version of the Type A Modified Nambu, with as many as 400,000 produced. Most Type 14s are marked with the month and year of production according to the year of the emperor. Most of the Type A Modified and Type B Nambu pistols were produced by the Tokyo Arsenal with a few pistols being produced by the Tokyo Gas & Electric Company. The Type 14 Nambu was produced by Nagoya Arsenal.[124]

Type 26 revolver. This was a double-action only, break top handgun, chambered for a proprietary 9mm round. Beginning in the 1870s, Japan adopted and purchased versions of the Smith & Wesson Russian and New Model 3 revolvers. Imperial forces continued to be supplied with the S&W new model number 3, ultimately ordering some 7,500 over the years. Chambered in .44 Russian, they were a top-break SA and relatively large pieces of hardware. In 1887, it was decided to produce an indigenous revolver, one that was smaller and of lighter caliber. Experiments were undertaken at the Imperial Japanese Army Tokyo Arsenal, and in 1894 the Type 26 Revolver was adopted, taking its designation from Emperor Meiji's 26th year on the throne. It was one year earlier than the gun's acceptance date, but the year in which the design was perfected.

The Type 26 was smaller than the S&W Russian, measuring 9¼ inches long as opposed to the S&W's overall length of 11½ inches. It weighed 8 ounces less that the S&W revolver. Like the Smith & Wesson, the Type 26 was a top-break revolver which featured automatic ejection of spent cartridges. It was chambered for a unique 9mm thin-rimmed

[124] *Japanese Military Cartridge Handguns 1893-1945.* Schiffer, 2003. See also Hogg, Ian and Weeks, John. *Pistols of the World.* London, DBI Press, 4th Edition, 2004.

smokeless power cartridge similar in dimensions to the .38 S&W with a bullet weight of 150 grains. Muzzle velocity was 640 feet-per-second with a muzzle energy of 135 foot-pounds. In comparison, the black powder .44 Russian cartridge was loaded with a 246-grain bullet that had a speed of 770 feet per second, which provided muzzle energy of 324-foot pounds.[125]

Type 04 Nambu. This was the original semi-automatic design, chambered in the under-powered 8 mm Nambu cartridge. It was adopted in 1904. The Nambu Type 14th Year Pistol, which appeared in 1925, was an improved form of the semi-automatic 8mm Nambu "4th Year Type" pistol (or "Model 04") which was officially accepted into service by the Imperial Japanese Navy in 1909 with a modified form appearing in 1915 (the 4th year of the Taisho era). The 4th Year Type, which showcased a design layout not unlike the famous German Luger, but an internally different pistol, saw combat actions in World War 1 but proved unreliable on the whole and expensive to produce for Japanese industry. It had a grip safety, but no magazine safety; also had an adjustable rear tangent sight. It was never actually officially adopted by the IJA. Officers were simply encouraged to buy their own, as the Japanese Army at the time did not standardize sidearms for COs and NCOs. The first models were known as the Type A, or "Grandpa Nambu." The Nambu 4th Year Type survived largely as a commercial venture until 1925.[126]

The Type 14 Nambu was a semi-automatic handgun was a modification of the 04, adopted in 1925. It had no grip safety, but was equipped with a magazine safety. The rear sight was fixed. There was a Type 14 Nambu, "Manchurian" modification. This is a Type 14 with an enlarged trigger guard to allow use with a gloved hand. It was designed with the goal of simplifying manufacturing in order to reduce cost. It was officially adopted for issue to non-commissioned officers in the Japanese Army in 1927 and was available for purchase by officers. The Type 14 was an improved version of the Type A Modified Nambu. [127]

The Nambu Type 19 "North China" pistol was produced with the

---

[125] Derby, Harry L.; Brown, James D. *Japanese Military Cartridge Handguns 1893–1945.* Philadelphia: Schiffer, 2003.

[126] Skennerton, Ian. *Japanese Service Pistols.* Queensland: Arms and Militaria Press, 2008. See also https://guns.fandom.com/wiki/Nambu_Type_4

[127] https://www.militaryfactory.com/smallarms/detail.asp?smallarms_id=475

reliability improvements seen on the Type 14. It was manufactured in occupied China. The quality of the handguns varied considerably. North China Type 19 pistol (not to be confused with the North China Type 19 rifle) is an improvement on the Type 14 Nambu pistol design which was manufactured in very small numbers in Japanese-occupied China late in World War Two. With shipping connections between Japanese troops in China and the Japanese home islands cut off by US naval activity, plans were put in place to expand Japanese-run arms manufacturing in China and the Type 19 pistol was part of those plans. The two major improvements to the Nambu design that the Type 19 incorporates are a solid frame with a disassembly lever in place of the removable trigger guard, and a redesign of the manual safety to be much more ergonomic. The magazine retention spring was also wisely discarded. Two distinct types of these pistols are known; one very well made with a bullseye style of final acceptance mark and one of very poor machining quality with a Japanese "2" character as a final proof, presumably indicating second quality. The explanation for these two distinct types is not known. The original intention of IJA was to produce 5,000 Type 19 pistols, but no more than about 200 were actually made. This is one of the scarcest Japanese WW2 pistols, and very interesting as an improvement to the Nambu design.

The Type B, 7mm pistol was the so-called "Baby Nambu." A very rare handgun, it was chambered for a 7mm proprietary Nambu cartridge. It was apparently originally designed for commercial sale, with most known going to high-ranking officers. This pistol was about three-quarters of the size of Type A. [128]

The Type 94 Nambu was a truly terrible handgun, intended originally for commercial sale, introduced in 1934, there is virtually nothing to recommend it. Nambu Type 94 was a pistol which required two hands to operate.[129] Because of a major design flaw it can simply kill or injure the user through a misfire by just moving around with it tucked in

[128] Van Lund, John C. *Japanese Military Handguns and Holsters, 1893-1945*. Blue & Grey Press, 1972. See also Skennerton, Ian. *Japanese Service Pistols Handbook*. Arms and Militaria Press, 2008.

[129] Skennerton, Ian, *Japanese Service Pistols Handbook*, p. 16; also, Kinard, Jeff. *Pistols: An Illustrated History of Their Impact,* p. 247.

the holster. The external sear could be compressed, causing the pistol to discharge. It survived largely because it was far cheaper to manufacture than others. Issue included tank crew members. A recent article on the world's poorest hand arms summed it up well:

> Japan has never been known as a firearms innovator, and many of its imperial-era small arms were inspired by or licensed from European designs. In 1934, Japanese firearms designer Kijiro Nambu unveiled the Type 94 pistol. The Type 94 was a blowback-operated pistol chambered in 8-millimeter. It weighed 1.68 pounds and was equipped with a six-round magazine. Not only was the Type 94 an ergonomic nightmare, it had the unfortunate tendency to discharge while being handled. Pressing an exposed sear on the side of the Type 94 will actually cause it to go off. It was also possible to depress the magazine release simply by laying the pistol down on a flat surface, and the pistol was known to be extremely difficult to field strip.[130]

Hino–Komuro pistol Type Sugiura .32 automatic Colt pistol caliber (7.65mm) pistol was manufactured in occupied China. Type Sugiura .25 automatic Colt pistol caliber (6.35mm) pistol was also made in occupied China. Although some policemen used the .32 ACP caliber pistols in Europe, this is certainly not a man-stopper and not suited for combat use. The .25 ACP caliber is even worse and was used by no other police or military.[131]

Hamada Type Pistol. The Hamada Pistol design was the work of Bunki Hamada. Prior to the commencement of all-out war, Hamada had worked in the sporting guns industry. It was essentially a Japanese version of the Browning .32 semi-automatic pistol.[132] In 1941, the Japanese government required the company to abandon civilian manufacture and support the war effort directly. The manufacturer began development of the design as a military weapon in late 1941. The Hamada design stole

---

130 Mizokami, Kyle. "These Guns Might Be Cursed: 5 Guns that Might Be the Worst Ever" *The National Interest,* 21 December 2019. See also https://www.thefirearmblog.com/blog/2013/10/31/worst-pistol-ever-type-94-nambu/

131 Brown, James D. *Imperial Japanese Handguns 1893-1945.* Schiffer Military History, 2007. See also Hogg, Ian, *Pistols of the World.* London: DBI Press, 4th Edition, 2004.

132 https://www.forgottenweapons.com/other-handguns/japanese-hamada/

from the proven Browning Model 1910 semi-automatic pistol. It was eventually chambered for two different cartridges: the 7.65mm [.32 automatic] Browning or the local 8x22mm Nambu pistol cartridge. Those made at the Nagoya Arsenal were slightly enlarged and re-chambered to use the 8x22mm Nambu pistol cartridge. Magazine capacity varied according to caliber. In 1943, a new striker component as well as a new frame plate lock arrangement were added. A key engineering change in the Hamada pistol was its switch to a dovetail joint at the barrel-to-frame connecting point; the original Browning relied on interrupted lugs. A recoil-operated (blow-back), locked breech action was used and overall weight reached 650 grams and overall length measured 6.5 "with a barrel assembly of 3.5" long. Sighting was through a V-notch and blade iron arrangement over the weapon. Compact and lightweight, the Hamada Type pistol was expected to directly challenge the troubled Nambu standard issue pistol of the Imperial Japanese military. The pistol was officially adopted into service by the Imperial Japanese Army with most issued to elements in China. Unlike other Japanese small arms of the time, particularly those of the late-war period which saw degraded production quality as the war wore on, the Hamada Type pistol managed to keep its relatively high standard of construction in place. Production of the pistol ended in 1944 with the exact figures unknown, but estimated at less than 5000.[133]

Machine Guns

The Japanese actually produced a rather decent personal machine gun. The first was the Type II sub-machine gun, which the Imperial Japanese Navy introduced in time for the invasion of Shanghai. Originally it was chambered for the 8mm Nambu pistol round, a wholly insufficient cartridge. Also, it had a tendency to jam. A smaller version of the gun was developed. Type II also had the unique distinction of being the first small machine pistol to have the magazine clip extend from the pistol grip. The gun was produced in small numbers.

---

[133] Smith, W. H. B. *The NRA Book of Small Arms: Pistols and Revolvers.* Washington, D.C.: National Rifle Association, 1946. See also McNab, Chris, *The Great Book of Guns.* Greenwich, 2004.

The comparatively more common sub-machine gun was the Type 100. It was also designed and produced by the Nambu Arms Manufacturing Company, and was introduced in 1942. The Japanese had studied copies of the German MP18 [also known as Swiss SIG Bergman 1920] machine pistol. The gun only entered service after the war had begun, and by the end of the war only 30,000 had been delivered. Compared to the number of MP-40s, British Sten Guns, Soviet PPSh-41s and American Thomas sub-machine guns, this number is relatively minuscule. Type 100 was an iconic looking firearm. It was simple, and inexpensive to produce. It was equipped with a bayonet lug. The Type 100 was an automatic-only, air-cooled, blow back gun that fired from an open bolt. As with the MP18, or the later Sten, it utilized a side-mounted 30-round detachable box magazine. The gun fired the grossly under-powered 8x22mm Nambu cartridge. Type 100 had a chrome plated bore, which was designed for jungle conditions. Some early models also featured a bi-pod, wholly impractical on such an arm. A paratrooper model was also fitted with a folding stock. The gun underwent a design change in 1944 to simplify the production. Both were variants on basic theme. Blow-back operated, box magazine Some 100/40s were made for paratroopers with a folding stock hinged through the wrist. Not many of either type issued. Type 100 sub-machine-gun simply did not function reliably, and only a few were ever made, probably about 300, and those for special forces. Also called machine-pistols in some European nations, these small usually fully automatic weapons were used in close combat and by many paratroopers. Most nations had such arms since World War I. United States used the Thompson sub-machine gun and also the Rising and M-3 "grease gun" automatic small arms.[134]

The Japanese produced several machine guns, none especially competitive with Western arms. With one exception, Japanese machine guns required the use of either oiled cartridges or an oiling mechanism in

---

[134] Ezell, Edward Clinton. *Small Arms of the World. The Basic Manual of Military Small Arms. American, Soviet, British, Czech, German, French, Belgian, Italian, Swiss, Japanese and All Other Important Nations.* Harrisburg: Stackpole, 1960. See also McNab, Chris, *The Great Book of Guns.* Greenwich, 2004. See also Bishop, Chris, and Drury, Ian. *Combat Guns: An Illustrated Encyclopedia of 20th Century Firearms*. Chartwell Books, 1987.

the gun to ensure that the action did not rip the head off the cartridge during extraction. Many Japanese light machine guns also had provisions for attachment of a bayonet.

Type 3 also known as the Model of 1914 and the Type 92 heavy machine gun were both copies of the French Hotchkiss machine gun. The Type 3 was chambered in 6.5mm Arisaka, the Type 92 in 7.7mm Arisaka.

The Type 11, introduced in 1922, was a squad-level gun chambered in 6.5 mm. One loaded it from 5-round Mauser stripper clips that were dropped into a hopper. Theoretically a good idea, as you can get ammo from riflemen if you run out, but rifle ammo proved to be too powerful. Instead of redesigning the gun to work with the standard rifle ammo, the entire concept was defeated by adopting a less powerful round. The Japanese Army had since the opening to the west followed French influence. This was certainly the case with the Type 11, which was modeled after the French Hotchkiss air-cooled, gas-operated light machine gun, and designed by Kijiro Nambu. It used the same cartridges as the Type 38 infantry rifle, and used a detachable hopper magazine, which allowed for the gun to be constantly fed with ammunition while firing. Five round clips could be stacked laying flat above the receiver, which eased loading – something that probably seemed like an excellent idea in test situations. This proved to be a problem as dirt and grime could easily jam the weapon. Likewise, while reloading from a fixed position was easy, it was nearly impossible to reload quickly while on the move.

The Type 92 utilized a slightly different gas system and different kind of grips. Other than that, they were very similar guns. Both fed from 30-round stamped steel strips. Type 92 machine gun was copy of Lewis gun that used the Japanese7.7 mm ammunition.

The Type 96 machine gun was chambered in 6.5 mm was designed to replace the Type 11, although it never really did so. This light machine gun was based on the Czech ZB which were captured from the Chinese. Type 96 looked quite a bit like the Czech ZB and the British Bren, but had few of the good features of those guns. The hopper mechanism replaced with a 30-round box magazine. Many had provisions for mounting the same type of telescopic sight found on the Type 97 sniping rifle. Type 96 was supposed to be an improved version of the Type 11. It featured a folding bipod and could even be fitted with a bayonet. The biggest problem with this light machine gun was that it jammed when fired cases often became stuck in the chamber. To solve the problem, it was suggested by the gun's designer Kijiro Nambu that the cartridges should be oiled, which in combat only made matters worse as debris adhered to the sticky cartridges jamming the weapons until the chamber was cleaned.[135]

The Type 97 light machine gun fired a 7.7 mm cartridge, similar to, but not interchangeable with, the 7.7 rifle cartridge. This was a version of Type 96 that was widely used on Japanese tanks. The Type 97 came into service in 1937, and was used on all Japanese tanks and other armored vehicles until the end of the war. The Imperial Japanese Navy also used the weapon in their combat vehicles such as the Type 92 Jyu-Sokosha Heavy Armored Car (mini-tank). It was much less common as a stand-alone infantry gun due to its significant weight. There was also a Type 97 aircraft machine gun that was the standard fixed light machine gun on aircraft of the Imperial Japanese Navy during World War II. Adding to the confusion of weapons, this weapon was not related to the Type 97 heavy tank machine gun used by the Imperial Japanese Army in ground combat.[136]

The Type 99 machine gun was essentially a Type 96 updated to 7.7 mm caliber. There was a major improvement over the Type 96, for a long last provisions were made for primary extraction of the fired cartridge, meaning that rounds no longer had to be oiled before use. Altogether the most successful and reliable of all Japan's machine guns. The caliber 7.7 mm was sufficiently powerful for use is a light machine gun, being quire comparable to the American .30 caliber 1906. Beginning with the few good features of the Type 96, while trying to address its flaws, the Japanese introduced the Type 99 in 1936. The gun is often compared to the British Bren Gun, but this is a useless comparison. Other than the top loading curved magazine and basic silhouette respective internal workings of these weapons are quite different. The Type 99 was actually

135 "The Japanese Type 96 Light Machine Gun – a Troubled Weapon." *Daily Kos.* 20 June 2013.

136 Mikesh, Robert C. *Japanese Aircraft Equipment 1940-1945.* Atglen, PA: Schiffer Military History, 2004.

an upgrade to the Type 96, and early models employed a mono-pod at the stock as well as a flash suppressor. One feature it did share with the Bren was that the barrel could be rapidly changed to avoid overheating. The distinctive feature, and what makes the comparison to the Bren inevitable, is that aforementioned top-mounted detachable box magazine, which held 30 rounds and solved the loading problems. As with the Type 11 and Type 96, the Type 99 was used throughout World War II.[137]

Anti-tank Rifle

As the Japanese Army became concerned following reports of Chinese purchases of British-made Vickers 6-Ton tanks, along with rising tensions about the Soviet Red Army along the Manchurian border, the IJA decided to create an anti-tank rifle in 1935. The Nagoya Arsenal submitted a weapon derived from their copy of the 13.2-millimeter Hotchkiss M1929 machine gun and the Kokura Arsenal submitted a new design using a 20 by 125 mm cartridge. Following testing in March 1936, neither gun was deemed satisfactory and both guns were returned to their respective designers to rectify problems encountered during the trials. The Kokura Arsenal built eight prototypes for the second round of trials held at the Army Infantry School in 1937, following which the IJA rejected the Nagoya weapon. IJA identified several problems to be remedied and a batch of fifty guns was built for additional trials in 1938, after which the weapon was accepted as the Type 97 Automatic Rifle.

The gun utilized a gas-operated delayed-blowback mechanism in which the barrel and receiver also recoiled to help steady the weapon. The weapon lacked a switch to change from single-shot to fully automatic mode. The Type 97 was the heaviest anti-tank rifle of World War, weighing 52 kilograms (115 pounds). It also had a gun shield which weighed 68 kilograms (150 pounds) plus four carrying handles. It employed a seven-round removable box magazine mounted above the receiver which fired at the rate of a dozen rounds per minute. The Type 97 fired solid-steel armor-piercing-tracer (AP-T), high-explosive-tracer and high-explosive incendiary-tracer shells. IJA credited it with the ability to penetrate 9 mm

---

[137] McCollum, Ian. "RIA: Nambu Type 96 & Type 99 LMGs". *Forgotten Weapons,* 3 September 2015.

(0.35 inches) of armor at 2,000 meters.[138]

The Type 97 was a bastard arm utilizing a conglomeration of ideas from various European guns. It was chambered fora special 20mm round not used in any other arm. It utilized a very complex operating system, employing a combination of gas operation and blow-back. Its action allowed a cartridge to recoil along a slide against heavy recoil springs and an oil buffer recoil system.[139]

---

[138] Williams, Anthony G. *Rapid Fire: The Development of Automatic Cannon, Heavy Machine Guns and Their Ammunition for Armies, Navies and Air Forces.* Shrewsbury, UK: Airlife, 2000; Zaloga, Steven J. *The Anti-Tank Rifle.* Oxford, UK: Osprey, 2018.

[139] Nafziger, G. F. *Japanese Infantry Organization and Tactics in World War II*: Volume II, *Night Attacks, Machine Gun & Infantry Gun Training, Battalion Training, Infiltration &Concealed Movement. . . .* West Chester, PA: Nafziger, 2004. See also Hogg, Ian, *Military Small Arms of the 20th Century.* London, 7th Edition, 2000. Sec also Kenda, Daniel J. *Lessons Learned from the Use of the Machine Gun During the Russo-Japanese War and the Application of Those Lessons by the Protagonists of World War I.* Biblioscholar, 2012.

Japanese Self-propelled Armament

Like its small-arms equivalent, Japanese tank design and construction fell far behind that of the Allies. While the tanks and their cousins were more than sufficient to overcome opposition in Korea and then Manchuria and finally China, they were never serious competition for American and Soviet designs. American tank crews tended to dismiss Japanese tanks as worthy opponents early in the Pacific War. Put another way, the Japanese air arms of both the Army and the Navy started the Pacific War with plans that were clearly superior to anything the Allies flew. But both Japanese forces had lost all claim to superiority by the end of the conflict. But Japanese tanks and related equipment began the Pacific War far behind those used by the United States and its allies and not only failed to match Allied pre-war quality but failed to produce a tank competitive with the basic Allied tank of choice, the U. S. M-4 Sherman.[140]

Type 89 Chi-Ro Light-Medium Tank

In 1925, the Imperial Japanese Army began development of its own medium-light tank. The initial product was judged to be too heavy and fitted with an unreliable engine. The Chi-Ro was of a conventional design and layout for this period, based heavily on the British Vickers Medium C light tank which had just recently been purchased. Operation required four crew members: commander, gunner, bow-gunner and driver. The bow-gunner and the driver maintained positions at the front hull while the commander and gunner resided in the forward-set traversing turret. The engine was fitted to the rear. The design maintained a pronounced nose-up appearance which worked well when traversing uneven terrains.

---

[140] See generally, Foss, Christopher. *Tanks: The 500.* Crestline, 2003; and Tomczyk, Andrzej. *Japanese Armor*. AJ Press, 2002.

The turret featured slanted armor plating for minimal ballistics protection as well as a circular commander's cupola. Main armament was a 57mm main gun with one hundred rounds of a new high explosive type. There was also a pair of 6.5mm machine guns with one facing the rear. Power was supplied by a Mitsubishi gasoline engine which was later replaced by a diesel engine. The Type 89's top speed was 15.5 miles per hour and even that under ideal conditions. It had a maximum range of 99 miles. With a weight of 12.8 tons the Army classified the Type 89 as a medium tank. The number 89 indicated the adoption in the Japanese calendar year of 2589 [1929] although production really began in 1931.[141]

Type 97 Medium Tank

During the mid-1930s, the Imperial Japanese Army worked on designs for the Type 97 Chi-Ha series of medium-class infantry support tanks. The tank weighed 15 tons and featured thin running tracks along each hull side. The tank required a crew of four. Armament included a 57 mm Type 96 low-velocity main gun and two 7.7 mm Type 97 machine guns. Armor protection varied on parts of the vehicle, ranging from as thin as 8 mm and as thick as 26 mm. Service entry occurred in 1938. Some 1,160 of these tanks were produced.

Following its introduction in East Asia, the Type 97 Medium Tank began to show its limitations. When compared with similar armed vehicles being deployed in Britain, the United States, Germany and the Soviet Union, the Type 97 was a failure. By 1941, the 57mm armament was severely outdated in its battlefield reach and penetration value. By 1942, it was severely outclassed though production was allowed to continue for lack of anything better. Additionally, the vehicle's thin armor protection made it vulnerable to both enemy tanks and to mobile anti-tank teams. A battlefield tank with pre-war beginnings was anything but a front-line tool at this stage in the war. The Army had envisioned the Type 97 as an infantry support mechanism, designed to work in conjunction with infantry forces. Its designers never conceived of it as an arm explicitly designed to take on enemy armor. The Army now concluded that a direct-

---

[141] Foss, Christopher. *Great Book of Tanks: The World's Most Important Tanks from World War I to the Present Day.* Zenith Press, 2003.

combat tank design was now in order to counter the growing use of superior Allied armor in the Pacific Theater.[142]

Type 1 Tank

The Army then looked for a new and improved medium class tank type, but also made along the same functional lines of the Type 97. Design work on the Type 1 began in 1940. The first vehicles appeared in 1941 with production delayed because of the lack of resources. From 1943 to 1944, manufacture was limited to just 170 vehicles The Type 1 *Chi-He* proved of relatively similar scope and function though more modern than its predecessor. Army added a 47mm high-velocity main gun which was selected and mounted to a centrally-located turret along a fixed hull superstructure roof. Power was supplied by the Mitsubishi 100 air-cooled V-12 diesel-fueled engine which turned out 240 horsepower. The new engine allowed road speeds of up to 27 miles per hour and an operational range of 130 miles. The 7.7 mm machine guns were retained, with one fitted co-axially in the turret and the other in the front left hull while the driver seated front-right. The Type 1 tank required five men to operate. Armor protection ranged from 8mm to 50mm and overall weight was increased to17 tons. This was the first Japanese tank to feature wireless communication; the previous communications were accomplished with signal flags. The Type 1 was a heavier design with improved armor protection and a more potent main gun armament than the Type 97, although still inferior to Allied tanks. A few Type 1 tanks are thought to have been fielded during the Battle of Leyte in the Philippines campaign. The Type 1 was largely retained for the defense of the Japanese homeland in anticipation of the Allied invasion. These tanks would not have fared any better than other Japanese light and medium tanks against the American M-4 Sherman. The M-4 outmatched the Type 1 design in every category. It was heavier, better armed, and fitted with superior armor; and it was evolving with every given month in the war

The Japanese arms industry never managed to produce any truly successful tank designs for its army in World War 2 and the Type 1 continued this trend. Japanese medium tanks, as classified by the

---

[142] Zaloga, Steven J. *Japanese Tanks 1939–45*. Osprey, 2007.

Imperial Japanese Army, were by Allied standards "light" tanks. That factor alone placed them at a distinct disadvantage on the modern battlefield.

The Type 1 *Ho-Ni* appeared in two other related battlefield forms, the Type 1 Ho-Ni II and the Type 3 Ho-Ni III. These vehicles were built atop the existing running gear and chassis of the Type 97 *Chi-Ha* Medium Tank.[143]

Tank Destroyer.

To fill the Imperial Japanese Army's need for a tank destroyer, engineers re-purposed the power pack and chassis of the medium tank and removed the turret structure. In place of the turret engineers installed a 75mm Type 90 field gun in a fixed, open structure. Only the front and sides of the enclosure were protected from the weather and attack. A fixed superstructure provided the necessary workspace for the gunnery crew but also served to restrict the traversal and elevation of the gun unit. In essence, the entire tank had to be turned towards the intended field of fire. Armor protection ranged from 25mm to 51mm and power was served through a Mitsubishi SA12200VD series V12 air-cooled diesel engine which propelled the vehicle to speeds of 24 miles per hour and ranges out to 120 miles. The vehicle was equipped with fifty-four 75mm projectiles and the operating crew numbered five. Beyond the main gun there was now defensive armament fitted save anything personal carried by the crew. The resultant design became known as the Type Ho-Ni I which appeared in limited numbers during 1942. The Type 1 Ho-Ni was deployed for the first time in the Battle of Luzon during the Philippines Campaign in early- and mid-1944. The Imperial Japanese Army lacked the numbers to affect the outcome of the battle positively and results shown the vehicles themselves were mixed. Only twenty-six first-series vehicles were produced.

The Type 1 Ho-Ni II was another Type 97 tank development but this vehicle emerged more as a self-propelled gun platform fitting the 105mm Type 91 howitzer instead of tank-killing armament. This

---

[143] Hara, Tomio. *Japanese Medium Tanks*. AFV Weapons Profiles No. 49. Profile Publications, 1972.

was set within a slightly revised armored superstructure and intended as a fire support vehicle.

A more closely related development of the Type 1 Ho-Ni I became the Type 3 Ho-Ni III which now fitted the more-powerful 75mm Type 3 anti-tank gun in a fully-enclosed armored superstructure that protected the crew. Total production numbered just 111 of these vehicles and most were saved to assist in repelling the anticipated invasion of the home islands.

The Type 1 tank design also appeared as a self-propelled anti-aircraft gun system. However, only a single prototype of the system was completed which was armed with two 37mm auto-cannons. That system could have been fielded as an anti-tank and anti-personnel system as well.

Near the end of the Pacific War, Japanese field commanders realized that nothing in the inventory of the Japanese Army would be able to withstand the increasingly advanced tanks and armored vehicles being implemented by the Allies. Thus, a more powerful version of the Type 3 Ho-Ni III was necessary. Development was rushed through on a new design, which was completed in 1944. The Japanese Army immediately issued an order for 200 units to be completed in 1945. The Type 5 Na-To made use of the chassis and superstructure of the Type 4 Chi-So armored medium tracked carrier. The superstructure had an open top and rear, with an enclosed armored drivers cab. For the Type 5 Na-To there was added a "shielded platform" for its main gun. Its main anti-tank armament consisted of a Type 5 75 mm tank gun, which was the same gun mounted in the Type 4 Chi-To medium tank. It was equipped with a variant of the Japanese Type 4 anti-aircraft gun in 75 mm. However, by that time production was impossible due to material shortages, and by the bombing of Japan in World War II, and testing was not yet completed by the end of the war. [144]

Medium Tank.

As American M-4 Sherman tanks were deployed in the Pacific War it became apparent the Imperial Japanese Army had nothing to compete with them. As we have seen, the so-called medium Japanese tank

---

[144] Tomczyk, Andrzej. *Japanese Armor* Vol. 5. AJ Press, 2007; and Zaloga, Steven J. *Japanese Tanks 1939-45*. Osprey, 2007.

was in reality a small or light tank suited only for use against secondary opponents such as the Chinese. Thus, the Japanese Army sorely needed to develop a true medium tank if it was going to be able to meet the Sherman head-on. It seems incredible that Japan would enter the war without a true medium-format tank of superior design and with sufficient firepower especially from its main gun.

The Type 4 *Chi-To* was one such development that was intended to help level the playing field. It was to be a true medium-class tank system, weighing in at 33 Short Tons and armed with a long-barrel Type 5 75mm main gun which was based on the proven Japanese anti-aircraft weapon. That it was a competent and useful design is not especially surprising because it was based on the original Swedish Bofors design. Armor protection was 75mm at the thickest point. The track systems featured seven small road wheels to a track side with the drive sprocket at the front, the track idler at the rear and three track return rollers. It was powered by a single Mitsubishi air-cooled V12 diesel-fueled supercharged engine putting out 412 horsepower. This provided the Type 4 with a top speed of 28 miles per hour and with an operational range of 160 miles. The 75mm main gun was backed by a pair of Type 97 Light Machine Guns for anti-personnel use. There were also mountings on the turret roof for adding another machine gun for close-in anti-aircraft defense. The turret itself sported a rather tall profile, clearly a negative factor on the battlefield. The turret was completed with slightly angled side surfaces and shaped as a hexagon when viewed in the top profile. The Type 4 tank had a crew of five: a driver, tank commander, gun layer, ammunition loader and radio-operator who also operated a machine gun.

Design and engineering of the Type 4 began in 1942 and continued through the years following. It was not until 1944 that the first completed Type 4 tank was made available. The Japanese explained that the slow development cycle and delays in reaching serial production were largely attributed to the Allied air campaign which targeted Japanese industry. As the Japan lost more and more of her possessions in the Pacific Theater, attacks then concentrated on the Japanese home islands. The highly successful air war struck at the heart of her industrial output, disrupting supply lines. General industrial production was disrupted and design initiatives such as the Type 4 all but halted. The new medium tank languished, with nothing more than two fully completed vehicles to show. At this point in the war, Type 4 tanks would have been used in defense of the Japanese mainland from the expected amphibious invasion from the Allies.[145]

As with any weapon that was never actually tested in combat situations, it is impossible to say how successfully the Type 4 might have been. On paper it should have been only slightly inferior to the Sherman tank. Still with the high profile it would have been more vulnerable than most medium tanks, especially to assault by aircraft.

Heavy Tank

The Imperial Japanese Army proposed the *O-I i*nitiative to create a super heavy tank. The initial plan called for a 120-ton tank to be armed with a 105mm main gun in a main turret with smaller turrets mounting smaller-caliber weaponry. The vehicle would require 11 personnel to manage its various functions. Another *O-I* plan was built around a 100-ton Ultra Heavy Tank. That design began with the conception of a mammoth vehicle showcasing as many as four turrets with large and small caliber armament. While these plans fell stillborn, the 120-ton endeavor did come to a certain degree of completion before the end of the Pacific War.

Outwardly, Type 120 O-I Super Heavy Tank exhibited a most conventional appearance, following the accepted straight-line-and-boxy-form design of previous Japanese armor. The vehicle was suspended atop eight road wheels to a track side, the tracks running the entire length of the hull and protected along its upper regions by armor skirting that was bolted onto the side of the upper hull. There was a short hull superstructure for the crew within and a main gun turret on top of the hull roof. A pair of smaller turrets were added ahead of the main turret in order to house additional armament. The main turret featured at least two access hatches along the roof with another at the rear facing. Additional fuel stores could be carried in drums along the rear of the hull as a vehicle this large was sure to consume the available internal stores rather quickly. The power-plant was set to the rear of the hull with access panels allowing for ease of

---

[145] Gander, Terry J. *Jane's Tanks of World War II.* Harper Collins, 1995.

maintenance of the large twin gasoline installations.[146]

Primary armament was a 105mm main gun which might have made an excellent mobile artillery system against Allied armor. This gun was to be based on an existing field artillery system, probably the five-ton Type 92, for it would have been a proven large-bore gun. Still, it would have required some modification to fit into a turret. The gun was to have been fitted into a fully-traversing turret with an overlapping rear section. The main armament would have been backed by secondary weapons in the form of a 37mm anti-tank cannon which was to have been fitted to a smaller traversing turret at the front hull, just ahead and under the main gun turret. Anti-infantry action would have been based on three 7.7mm Type 97 series machine guns. Ammunition supply would have included sixty projectiles for the 105mm gun, a hundred projectiles for the 37 cannon and as many as 7,470 rounds of 7.7mm ammunition. Power was to be supplied by a pair of V12 series gasoline-fueled engines rated at 500 horsepower each. The engines would be conventionally installed in a rear compartment of the armored hull. Armor would have had thickness of up to 200mm. Overall weight was in the vicinity of 120 tons, hence the designation.[147]

While the British and the Soviets had considered such a multi-turret combat vehicle, actions in World War II had shown the limitations of the multi-turret concept. Thus, the single turret became the norm, a standard which continues to this day. Nevertheless, the Japanese designers must have seen some value in the multi-turret tank concept. Since the tank was never produced in number, if at all, one can only speculate on its effectiveness. However, the plodding nature of the tank required that it be supported with infantry protection and medium and light tank support to avoid being overrun by enemy personnel. Such a huge tank would have made for a large and slow target on the battlefield. It would have been particularly vulnerable to low flying attack aircraft overhead.[148]

---

[146] Estes, Kenneth. *Super-heavy Tanks of World War II.* Osprey, 2014. See also https://www.militaryfactory.com/armor/detail.asp?armor_id=528

[147] Ibid. See also https://en.wikipedia.org/wiki/O-I_super-heavy_tank

[148] http://ftr.wot-news.com/2013/11/19/superheavy-japanese-tanks/. See also https://tanks-encyclopedia.com/ww2/jap/ww2_IJN_Tanks.php. See also Rottman, Gordon L. and Takizawa, Akira. *World War II Japanese Tank Tactics.* Osprey, 2008.

Type 3 Ka-Chi Amphibious Medium Tank

As early as 1928, the Japanese Army had been developing amphibious tanks and created several experimental models such as the Sumida Amphibious Armored Car. The very first amphibious tank existed only as a concept vehicle, with none produced for military use. The Special Type 2 Launch Ka-Mi was the first amphibious tank of the Imperial Japanese Navy. The Type 2 Ka-Mi was based on the Imperial Japanese Army's Type 95 Ha-Go light tank with major modifications. In 1940, the Imperial Japanese Navy took over development of amphibious vehicles with Mitsubishi as primary contractor. That firm used the knowledge gained from the SR program and the Type 95 Ha-Go light tank as the basis for development. The largely experimental Type 1 led to the development of the Type 2 Ka-Mi which was produced in 1942. The Type 2 Ka-Mi was designed for the IJN's Special Naval Landing Forces for the amphibious invasion of Pacific Islands without adequate port facilities, and for various special operations missions. However, only approximately 182 were ever built.

The Imperial Japanese Navy developed the Type 3 so that it could land troops and supplies on various islands which it had conquered. It had a 37mm turret gun and was operated by a crew of five. Supposedly some 141 vessels were manufactured and a few used in various island campaigns in 1941.

The Type 3 Ka-Chi was a heavily modified version of the chassis of the army's Type 1 Chi-He medium tank. It featured considerably better armor and far greater firepower than the earlier Type 2 Ka-Mi. It had smooth sides that faired into front and rear flotation pontoons made of sheet-metal. The front pontoon had a curved 'bow' shape and both pontoons could be jettisoned from inside the tank once the tank had landed. As it was, in practice, troops usually retained the pontoons because they provided a small amount of additional protection against enemy fire. Propulsion was provided by twin-screws. It had two steering screws. It had a distinctive large snorkel behind the turret for aerating the diesel engine more efficiently and keeping the exhaust free of water. The main gun was the same 47 mm gun used on the army's Type 97-Shinhoto Chi-Ha. Secondary armament was a coaxial Type 97 heavy tank machine gun. The vehicle required a crew of seven. The Type 3 Ka-Chi was produced in very limited numbers, but proved useful due to its capability to be launched from a submarine, which enabled it to accommodate the increasingly difficult task of daytime reinforcement for isolated island garrisons located in the South Pacific.[149]

Small Tank

By the 1930s major powers had thought of modern warfare as being based heavily on a fast-moving, mechanized army which would face with the enemy through both firepower and speed. One major component of this lightning warfare was the small tank. These battle tanks would spearhead an aggressive movement with infantry support following. Much infantry support would come from small tanks. Tanks, coupled with calculated air support, could turn any army into a major military power overnight. The Empire of Japan was quietly building up a war machine to conquer her interests throughout Asia and, ultimately, the Pacific and the light tank would, therefore, prove vital in the Empire's upcoming actions.[150]

The Type 95 *Ha-Go* was the light tank design primarily utilized by the Imperial Japanese Army during World War II. It was first fielded during the Second Sino-Japanese War but clearly made her mark against both British and American forces across Southeast Asia and in the Pacific theater. This light tank maintained an excellent top speed and was appropriately armed by mid-to-late 1930s standards. It operated with a crew of three and was utilized to good effect during the early campaigns of World War II. The Type 95 was the first enemy tank to be engaged by the Americans in the conflict. It was a light tank used by the Empire of Japan during the Second Sino-Japanese War, at Nomonhan against the Soviet Union, and in the Second World War. It proved sufficient against infantry but, like the American M3 Stuart light tank, was not designed to combat other tanks.[151]

---

[149] https://en.wikipedia.org/wiki/Type_3_Ka-Chi. See also Zaloga, Steven J. *Japanese Tanks 1939–45.* Osprey, 2007.

[150] https://en.wikipedia.org/wiki/Type_3_Ka-Chi

[151] Hara, Tomio. *Japanese Combat Cars, Light Tanks, and Tankettes*. AFV Weapons Profile No. 54. Profile Publications, 1973.

Based upon knowledge garnered from the procurement and operation of various foreign tanks, the Japanese produced their own light tank. They chose a design built on great speed but having only light armor protection. The Type 95 was unveiled the system in 1934 at the Sagami Arsenal. There were early concerns about the type's lack of armor protection for the crew and engine but this was deemed acceptable to maintain the required speed for the class. As such, the new tank design was assigned the rather plain designation of "Type 95" (nickname of "Ha-Go") and production began in 1935 at the Mitsubishi Heavy Industries.

Some 2,103 examples were ultimately delivered. Mitsubishi Heavy Industries was an early primary maker of the tank system though Sagami Arsenal, Hitachi Industries, Niigata Tekkosho, Kobe Seikosho and Kokura Arsenal combined to deliver over 1,200 additional examples by war's end. It should be noted, however, that critical war materials and technological advancements were primarily earmarked for consumption by the Imperial Japanese Navy over that of the Japanese Army. This often led to "second-rate" weapons developments for her land forces and can put into perspective the somewhat limited reach of the IJA during World War 2. Her tanks were generally outclassed by competing Allied designs, particularly late in the war, and her armored force was nothing in comparison to that of the Germans. Warships proved the top priority for the Japanese Empire in World War 2 and her land army would suffer for it.[152]

The Type 95 was made operational with the IJA in 1935. At its inception, the tank system was one of the better designs in the class of light tanks anywhere in the world. Its 37mm main gun was potent for its time and, as a whole, the Type 95 compared favorably to foreign designs. She was fast, well-armed for her role and helped to make those early IJA mechanized operations a success.

Success of the tank soon brought about several notable variants to her family line. The Type 3 Ke-Ri was a proposed light tank design fielding a heftier 57mm main gun though this tank was never put into production. The Type 4 Ke-Nu was produced in 100 examples and melded the turret of the Type 97 Medium tank with the Type 95 chassis for improved crew conditions. The Type 95 "Manshu" served as a Type 95

[152] https://www.militaryfactory.com/armor/ww2-japanese-tanks.asp

crew trainer. The Type 95 "Ta-Se" became a proposed anti-aircraft design fitted with a 20mm cannon. The Type 95 Ri-Ki was a battlefield engineer vehicle fitting a 3-ton boom crane. The "Ho-To" was a self-propelled gun mounting a potent 120mm main gun. The Type 5 Ho-Ru was another self-propelled gun initiative, this mounting a 47mm main gun. The Type 98 Ke-Ni was a lighter form of the base Type 95 yet fielded thicker armor. Some 200 examples of this improved tank were produced beginning in 1942.

The most notable Type 95 variant became the Type 2 "Ka-Mi" amphibious tank. This model was based on the Type 95 chassis but modified for use in amphibious operations and used extensively across the "island hopping" campaigns dotting the Pacific Theater.

Design of the Type 95 was conventional by any standard. Thin treads were placed on either track side. Drive sprockets were fitted forward with the idler at the rear. Two track return rollers were set under the top of the treads. There were four road wheels to a track side, these joined in pairs and utilizing a simple bell crank suspension system. The Type 95 took on a noticeable "nose up" appearance with the raised front hull and track system. The superstructure featured flat-faced and angled armoring across all facings. The forward facing just above the glacis plate sported a machine gun port on the left side and an access hatch with vision port on the right. The power-plant was fitted to the extreme rear of the hull with the exhaust system and muffler arranged along the right side of the rear hull. The rear hull was a large rectangular vertical facing. The turret, containing the primary armament, was cylindrical in nature with a hatch located at the top. The keen observer would note that it sat offset to the left of the hull superstructure. It maintained 45-degree rotation (manually-operated by hand) and was armed with the main gun and a secondary machine gun, the latter facing aft to cover the tank from the rear. The interior was lined with asbestos as a helpful layer in reducing heat buildup inside the tank as well as serving to pad the crew during travel on uneven terrain. Crew conditions were generally cramped, particularly in the turret and driving with hatches and panels open was done whenever possible, especially when combating the Pacific heat.[153]

---

[153] Tomczyk, Andrzej. *Japanese Armor*. 5 Vols. AJ Press, 2002-07; Zaloga, Steven J.

Japanese Tanks in the Pacific War

While the Japanese tanks scored a series essentially uninterrupted victories against non-existent Chinese armor, combat with allied forces was a wholly different situation. Even the destruction of a single armored vehicle could be counted as a Japanese victory so scare were successes. While most fighting in the Pacific took place primarily in tropical jungles and rain forests where tanks were useless there were a few encounters involving armored vehicles. Below is a simple summary of tank action in the South Pacific.[154]

Guadalcanal. In October 1942 there was a brief battle between Japanese and American light tanks. The Japanese First Independent Tank Company was converted from the Fourth Company of the Second Tank Regiment after the occupation of the Dutch East Indies. Twelve tanks were dispatched to Guadalcanal. Two of the tanks were damaged after they landed on Guadalcanal. The remaining ten tanks participated in the battle of Matanikau River on 23 October. On that date, two battalions of the Fourth Infantry Regiment and the nine tanks of the First Independent Tank Company launched attacks on the U. S. Marine defenses at the mouth of the Matanikau. U. S. Marine artillery, cannon, and small arms fire repulsed the attacks, destroying all the tanks and killing many of the Japanese soldiers while suffering only light casualties. The victors, American M-3 light tanks, were clearly superior to anything the Japanese had to send to the front.

Biak. The island of Biak dominates the entrance to Geelvink Bay, near the western end of New Guinea. The island was held by 11,000 Japanese troops under the command of Colonel Kuzume Naoyuki. He decided instead to allow the Americans to come ashore unopposed so that they would enter into a trap he had prepared for them. Naoyuki had a single company of nine Type 95 tanks. U. S. Army landed on 27 May 1944 with twelve Sherman tanks, 29 field guns, and 12,000 troops. Naoyuki's tanks attacked American Sherman tanks on the beach but were annihilated in a brief skirmish. On the morning of 29 May 1944 in Biak[155], three American

---

Japanese Tanks 1939–45. Osprey. 2007.

[154] Tomczyk. *Japanese Armor.* 5 vols. AJ Press, 2003-07. See vol. 3.

[155] Biak is a small island located in Cenderawasih Bay near the northern coast of

M-4 Sherman tanks advanced on the road to the airfield. Four Japanese light tanks countered them, but they were all quickly destroyed. Three additional Japanese tanks attacked but they too were rapidly destroyed. Tank Commander Iwasa was killed in this fight. The remaining two tanks were destroyed when the Americans attacked the airfield on 15 June.

Saipan. On 15 June 1944 some 8,000 Marines landed on the west coast of Saipan. Strategically placed Japanese artillery allowed the Japanese to destroy about 20 amphibious tanks, although none were damaged by Japanese tanks. While the Japanese defenses included the Ninth Tank Regiment of the First Tank Division of the Imperial Army, it was not a factor in the Saipan campaign, destroying no American armored vehicles. On the afternoon of 15 June 1944 on Saipan, the 4th company of the Ninth Tank Regiment made a counterattack against the U. S. Sixth Marine Regiment. The U. S. Marines landed along the Saipan beach near Charan Kanoa. Japanese tanks then rushed to the beach to prevent a successful landing. The Marines destroyed all the tanks with Bazookas. Another Japanese counterattack failed with the loss of all their tanks. The next day the Japanese undertook a large counterattack with all their forces. About 30 tanks in the van of Japanese forces began to attack after the midnight and rushed at the beachhead. The Marines concentrated their firepower on them and defeated Japanese forces completely, destroying most of their tanks. The next morning only a few Japanese tanks could withdraw from the beachhead. With this action the Ninth Tank Regiment on Saipan had essentially ceased to exist.

Guam. On Guam the night of July 21th, Japanese troops accompanied with 5 tanks of First Company of the Ninth Tank Regiment attacked American beachhead in the south of Agot, but they were repulsed by the American Marines. The Marines with their Bazookas and Sherman tanks destroyed all opposing tanks. Then the Second Company of the Ninth Tank Regiment and the 29th Division Tank Unit were ordered to participate in the general attack of July 25th. However, they lost their way and were not a factor in the attack. The remaining tanks fought delaying action and were destroyed one by one. Lieutanant General Takeshi

---

Papua, an Indonesian province, and is just northwest of New Guinea. Biak is the largest island in its small archipelago.

Takashima was killed on 28 July, so Lieutenant General Hideyoshi Obata assumed the command of the Japanese defenders. The Japanese counterattacks against the American beachheads, as well as the fierce fighting, had exhausted the Japanese. At the start of August, they were running out of food and ammunition and had only a handful of tanks left. Obata withdrew his troops from the south of Guam, planning to make a stand in the mountainous central and northern part of the island where tank battles were virtually impossible. On 10 August the last ten Japanese tanks attacked the American tanks and were utterly annihilated.

Tinian. In July 1944 American Marines landed and were opposed by six Japanese tanks. These armored vehicles made a counter-attack accompanied with infantrymen. Five tanks were destroyed in the resulting action. The remaining tank, along with others of the tank unit, was lost during the subsequent combat.

Sarmi, New Guinea. In July 1944, the Japanese scored one of the very few tank victories in the Pacific War. A tank unit was deployed in the rear near Sarmi. Two Type 95 Light Tanks were posted at the airfield in Wiske as an outpost. On 3 July, they encountered two American amphibious tanks, which were probing the airfield. One Type 95 Light Tank fired and destroyed one American amphibious tank. The other amphibious tank retreated. Later military action occurred mostly in the jungle so the Japanese tanks were not used.

Peleliu. The Japanese were well armed with mortars, backed by a light tank unit and an anti-aircraft detachment. On 15 September, sixteen Japanese tanks, accompanied with one infantry company, rushed at the landing beach. The 5th Marines made significant progress on the first day. They pushed toward the airfield, but were met with Colonel Nino Nakagawa's counterattack. His armored tank company raced across the airfield to push the Marines back, but the Japanese tanks were soon engaged by U. S. tanks, howitzers, naval guns and dive bombers. Nakagawa's tanks and escorting infantrymen were quickly destroyed. The Marines repulsed this attack who had been prepared for defense.[156]

---

[156] Takizawa, Akira and Rottman, Gordon L. *World War II Japanese Tank Tactics*. Osprey Publishing, 2008. See also *Japanese Tank and Antitank Warfare*. Military Intelligence Divison, 2013.

Japanese Air Force and Navy

The Japanese had fielded their Oscar and Zero fighters at the beginning of the Pacific War. These two aircraft were better than what American had at first; indeed, these two fighters were arguably the best then in existence. Whereas the U. S. quickly developed increasingly better airplanes and ships that soon surpassed the quality of the Japanese models, the Japanese quickly fell behind by standing pat with what were now obsolete airplanes. Every other major combatant continuously improved their aircraft, deploying improved versions of older models and developing new models. A highly perceptive article by Osamu Tagaya[157] provides us with the necessary overview of the strengths and weaknesses, successes and failures, of the air forces of the Imperial Japanese Navy (IJN) and of the Japanese Army, in World War II, in what they called the Pacific War.

Long before starting the Great Pacific War the Japanese leadership was well aware of Japan's limited capabilities in both resources and industrial capacity. Often quoted is Admiral Yamamoto's observation, based on his years of travel and study in the United States, that its industrial power was awesome. Fighting Third World powers such as the Chinese required far less industrial output than taking on a major industrial powerhouse such as the United States.

As a nation preparing for a major war, Japan failed to unify the two branches of its armed forces, the Imperial Japanese Army and the Imperial Japanese Navy (IJN). This unification was necessary in order to focus on a single strategy in the years leading to the war. The Army as the land power saw Soviet Russia as its great enemy, and thus the object of its war game planning. Conversely, the IJN focused on the United States as

---

[157]Tagaya, Osamu: "The Imperial Japanese Air Forces," in: Higham, Robin & Harris, Stephen, eds. *Why Air Forces Fail*. University of Kentucky, 2006.

its principal foe. The reason that the two branches could not agree was that neither was not subordinate to the civilian Japanese government. Thus, there was no unifying power to force the necessary cooperation. There were therefore two very different strategies competing for precious and limited resources, manpower and equipment.[158]

This also led to the parallel development of various aircraft and prevented the creation of the uniform standards that would have maximized resources. Such inefficiency in the use of resources is ordinarily problematic, but for Japan, in combination with limited industrial capacity and resources, such effects were damning.

The Japanese Army Air Force (JAAF) was primarily developed for tactical support of a land war against Soviet Union, and to a lesser degree against China. The JAAF had no use for, and very little interest in, naval navigation or long-range capabilities. These were crucial for IJN's use in South East Asia and especially the islands of the Pacific. As a result, the Navy had to fly long-range bombing missions in the Philippines for the Army.

The Japanese Army lacked strong advocates for air power in its ranks, primarily because the JAAF had a very limited number of army officers with air combat experience. Although the Japanese were among the first to use combat aircraft in World War 1, the time and number of such engagements was severely limited. Deployment of air power was far too limited to have the impact necessary in order to convince enough officers of the importance air power. The army's half-hearted attempt to create a wholly independent air force as a third branch of the armed forces, such as was occurring in Germany and the United Kingdom, met with complete failure. Many in the Navy feared that, should the independent air force be created, the IJN's Fleet Air Arm would be reduced to a marginal role at best.

Thus, initially, the main roles for the Army Air Force lay in recon and air combat. The JAAF's use of bombing missions received only

---

[158] Mikesh, Robert C. *Japanese Aircraft Equipment 1940-1945*. Schiffer Military History, 2004. Probably the definitive work on its subject, this book contains a wealth of information about Japanese aircraft equipment, from instrumentation through guns, gunsights, and bombsights.

limited attention. There was a consequent slow development of bomber squadrons in the 1920s.

When, for a short time in the late 1920s the Army joined the Navy in focusing on the United States as the main opponent, Japan began the development of a large four engine bomber. The two armed forces began to develop aircraft catapults to be deployed on land-bases in order to circumvent the problem of building long air strips. This was done in anticipation of an invasion of the Philippines.

Soon the Army focused again on Russia and on mainland Asia, and these projects were discontinued in the early 1930s. When Japan invaded and annexed Manchuria, the Japanese were faced with an extended land border to the Soviet Union. That changed the Army's strategic situation. Additionally, the Soviet Union had developed the TB-3 bomber which placed the Japanese home islands into the range of the Soviet Air Force.

In the mid-1930s, the Army undertook a major expansion of its air arm and, in 1937, declared the destruction of the enemy's air force as its primary mission. In the ongoing war in China and with the brief clashes with Soviet Union the JAAF mainly contributed on a tactical level. By 1940 the emphasis on destruction of the enemy air force was weakened and the offensive power remained mostly on a tactical level.[159]

The main problem with JAAF aviation lay in a lack of initiative by a conservative senior leadership that was interested in advancement of air power. JAAF primarily reacted to international developments rather than formulating its own doctrines. There seemed to be few chances for recognition and advancement in JAAF and this led to a shortage of officers with proper experience. JAAF was characterized by its poor leadership and unimaginative staff work, giving rise to operations that where questionable in their effectiveness and all too predictable and conventional in nature.

On the other side, the IJN was characterized by its strong advocates for air power. The Imperial Japanese Navy was a pioneer in use and development of naval aviation. It built the world's first aircraft carrier

---

159 Peattie, Mark R. *Sunburst: The Rise of Japanese Naval Air Power, 1909–1941*. Annapolis, Maryland: Naval Institute Press, 2013.

the *HIJMS Hōshō* in 1922. Too, it introduced the first operational deployment of an all-metal monoplane carrier-fighter plane in 1937. And in 1940 it was able to demonstrate the mass deployment of torpedo and dive-bombers in coordination with fighter aircraft. IJN successfully launched aircraft simultaneously from several aircraft carriers. At that time, it was something no navy was able to do. On the eve of the attack on Pearl Harbor, the IJN had more aircraft carriers than any other navy in the world. It also had the world's leading naval air arm with an elite corps of highly trained and motivated pilots.

IJN's Zero fighter outmatched all of it counterparts, whether Allied or Axis design and manufacture. Many experts, now and then, considered it to be the very best airplane for its purpose yet manufactured and deployed. On 28 March 1944, the U. S. Flight Test Engineering Branch concluded after testing a captured Mitsubishi Zero, "The airplane is highly maneuverable, has a fair rate of climb, and good visibility; however, its speed in level flight is low, it is lightly armed, has no armor protection for the pilot, and the fuel tanks are not self-sealing. The cockpit layout is fair, leg-room is insufficient for an average sized man ..." The Zero had abundant good qualities; it was reliable, had an extraordinarily long range, and was, above all, maneuverable and easy to fly. Even with the swathe of more advanced US fighters now arriving in the South Pacific, it was not wise to get into a prolonged dogfight with a Zero. [160]

The Oscar was second most important Japanese fighter airplane of the Pacific War. Technically, it was the Nakajima Ki-43 Hayabusa (meaning "Peregrine Falcon"), and was also called the "Army Type 1 Fighter." The Oscar was a single-engine land-based tactical fighter used by the Imperial Japanese Army Air Force in World War II. It bore a certain resemblance to the Mitsubishi A6M Zero. Both aircraft had generally similar layout and lines, and also used essentially the same Nakajima Sakae radial engine. Allied aviators frequently made mistakes in enemy aircraft identification in the heat of battle, erroneously reporting having fought Zeros where there were no Navy fighters. Like the Mitsubishi-produced A6M Zero, the Ki-43 was light and easy to fly and became

[160] https:// japaneseaircraftofwwii.blogspot.com / 2016/ 08 / degraded-performance-of-japanese-navy.html

legendary for its combat performance in East Asia in the early years of the war. In the early years the Oscar could outmaneuver any opponent. Like the Zero the Oscar did not have armor or self-sealing tanks, and early on its armament was poor. Allied pilots often reported that the nimble Ki-43s were difficult targets but when hit they burned easily or broke apart after only a few hits. The Ki-43 shot down more Allied aircraft than any other Japanese fighter. Produced as late as 1945, Japan made at least 5,919 of the aircraft.[161]

Furthermore, the IJN possessed a strong land-based naval bomber force, the so called *rikko* units. These units were initially developed in order to counter the limits on the number of carriers and carrier-based aircraft allowed under various Naval Treaties. It was aircraft of these units that sank the Royal Navy's battleships the HMS Prince of Wales and *HMS Repulse* early in the War.

IJN still harbored a strong old guard that preferred the firepower of battleships to the air power of carriers. IJN never did abandon the battleship as a core weapon prior to or after these initial successes of carriers. The main fleet organization still focused on the battleship. These powerful officers never did allow the creation of a complete carrier task force organization. Because of the powerful influence of the old guard traditionalists an enormous amount of resources went into building the *Yamato* and *Musashi* the super-battleships. This emphasis on traditional dreadnoughts is well illustrated by the presence of Japanese battleships at the Battle of Midway.

Another major flaw of the IJN was its habit of emphasizing attacks on enemy warships while ignoring enemy supply ships. For example, IJN ignored transport ships following its defeat of the Allied cruisers at the Battle of the Savo Island.

The focus of the Japanese Army and the Japanese Navy was on supporting specific engagements while neglecting logistics. The Japanese lacked pilots for ferrying aircraft to the front lines. The Japanese periodically used combat pilot to ferry planes due to lack of ferry pilots. Moreover, the Japanese could not rely on being able to use an existing

---

[161]Bueschel, Richard M. *Nakajima Ki-43 Hayabusa I-III in Japanese Army Air Force.* Reading, Berkshire, UK: Osprey Publications, 1970.

infrastructure in the Pacific since essentially there was none. Both the Army and IJN on occasion had to use naval bombers to drop supplies, because there were no other transport aircraft available. This was in stark contrast to the Allies that airlifted an infantry division from Australia to New Guinea.

Too, their capability of constructing airfields was severely limited. They lacked sufficient mechanized engineering equipment to create and improve existing airfields. It not only resulted in a huge delay and back-breaking labor on the Japanese side, but also the resulting installations were often very limited in size. Japanese airfields were usually congested with planes that were parked closely to each other. This congestion lead to severe losses when Allied airplanes strafed those air fields.

Both military units ignored warning and communication equipment and failed to create good radar and effective radios for fighters. Frequently, attacks occurred without IJN or the Army receiving any advanced warning. Most of the Japanese lacked any kind of radar equipment. Initially, the Japanese had been among the leaders in radar technology, but they fell behind by not investing in the technology for military purposes.

There was a lack of effective shortwave radios. Japanese fighter pilots communicated with visual signals. They never could fully exploit their initial advantages in training and equipment. Furthermore, it also prevented the creation of proper ground or carrier-based-control capabilities.

The lack of unified strategy between the Navy and Army showed a lack of foresight as well as strategic perspective. This failure prevented a proper and effective allocation of the nation's limited resources, both before and during the war. The situation became absolutely critical as Allied efforts to staunch up the flow of materials from Japan's conquered territories became ever more effective.

The Japanese aircraft industry lagged behind in the development of powerful engines. The Japanese solved the problem by reducing the weight of their aircraft. Japanese airplanes used little or no armor plates. Aircraft also lacked self-sealing fuel tanks. Misinterpreting their experiences fighting the Chinese, the Japanese concluded that these measures would have no negative effect on the successes of their aircraft. During the Guadalcanal campaign Japanese losses increased and their highly-trained airmen thinned out quickly in some large measure because of these two omissions. Tagaya wrote, "In the end, the initial margin of superior training and experience exhibited by its airmen proved insufficient to prevent serious attrition."

More than any other mistake, the Japanese failed to continue training a new generation of pilots. The initial assumption that it was better to have a small number of highly skilled pilots than a larger number of less trained pilots proved grossly incorrect. The serious attrition of which Tagaya wrote could have been overcome by continually training more and more pilots. The United States produced a number of pilots who were as accomplished as any group anywhere in the world; but it also continually trained more and more pilots.

In virtually all ways, Japan before and shortly after the First World War was essentially dependent on Western technology and imported most of its aircraft and equipment at that time. During the 1930s they reached a certain plateau in aircraft design, but their development cycles were quite long. In aircraft components and subsystems, such as radios, Japan was dependent on Western imports.

Prior to April 1942, there was no serious effort made to create a successor for the Zero fighter plane. Had such measures been taken in 1940, the Japanese might have had an aircraft to counter the Corsair or Hellcat. United States had tested the Zero and had developed fighters expressly designed to combat the Zero. Instructors taught tactics designed to take advantage of the Zero's shortcomings. But Japanese pilots had to fight with their modified version of the Zero.

By June 1944 the Japanese pilots faced highly trained American pilots who were equipped with new and vastly superior planes. Increasingly the IJN's carriers had only poorly-trained pilots. This resulted in an ineffective air force.

Decline of the Japanese Navy Air Force

At the beginning of the Pacific War, Japanese Naval and Military Air Forces had overwhelmed the Allies throughout the Asia-Pacific

Region. Japanese pilots frequently scored victories in ratios as high as ten to one 10:1 or more. The Brewster F2A Buffalo was an American fighter aircraft which most frequently saw service early in World War II. Designed and built by the Brewster Aeronautical Corporation, it was one of the first U.S. monoplanes specifically fabricated and developed for deployment on aircraft carriers. Although largely forgotten today, the Buffalo actually won out in competition against the Grumman F4F Wildcat in 1939. Although superior to the Grumman F3F biplane it replaced, and the early F4Fs, the Buffalo was largely obsolete when the United States entered the Pacific War, especially when compared to the Japanese Mitsubishi A6M Zero. Adopted by several nations, the Buffalo was most successful in Finland where it compiled a 36 to 1 kill ratio over planes from the aggressor Soviet Union in the Winter War. Only 509 Buffaloes were ever produced, yet the aircraft was piloted by some 40 air-aces, the highest ratio of aces per number of aircraft produced, of any production fighter plane. Outclassed by the Zero, its last major engagement was in the defense of Wake Island in June 1942. It was the only American fighter to have been shot down more frequently that its score of victories.

By the first half of 1943 the Commander of Air Forces in the Solomon Island chain (COMAIRSOL) had achieved a startling turnaround in performance. Beginning in the middle of June 1943 there was another huge upward surge in comparative performance of Allied fighter forces. By 16 June 1943 aerial photographs showed that Japan had increased aircraft at Rabaul to 245 and the subsidiary airfields, which ringed the main airfield, were also full of additional planes.

In Tokyo the developing catastrophe in the air was being hidden from senior commanders. For example on 14 April 1943 Admiral Yamamoto ordered a two-pronged force, code-named Y-1 and Y-2, consisting of 75 fighters and 23 dive-bombers from the Third Fleet along with 54 fighters and 44 medium bombers from the 11th Air Fleet, to make a major attack on Milne Bay, which had become an important logistical center for the Allied advance in New Guinea and the Solomons. In that engagement, Japanese pilots claimed to have shot down forty-four Allied aircraft. In fact, Allied losses amounted to a single P-40 and its pilot killed; four others were shot up and a P-38 crash-landed. Similarly exaggerated claims were made for ships sunk. IJN also falsely reported that it had sunk four U. S. transports and had heavily damaged six others. The metal for Japanese fighter planes was so inferior that Hashimoto heard that the engines cracked if they were flown at full throttle. This was rapidly becoming irrelevant as the country ran out of aviation fuel.

Allied aircraft under the command of Lieutenant-General Kenney scored a remarkable victory at the Battle of the Bismarck Sea on 2 to 4 March 1943.[162] That battle was a disaster for the Japanese. Out of 6,900 troops who were badly needed in New Guinea, only about 1,200 made it to Lae. Another 2,700 were saved by destroyers and submarines and returned to Rabaul. About 2,890 Japanese soldiers and sailors were killed. The Allies lost 13 aircrew, 10 of whom were lost in combat while three others died in an accident. There were also eight wounded. Allied aircraft losses were one B-17 and three P-38s in combat, and one B-25 and one Beaufighter in accidents. General MacArthur issued a communiqué on 7 March which claimed American and Australian aircraft had sunk 22 ships, including twelve transports, three cruisers and seven destroyers, had been sunk along with 12,792 troops.[163]

This build-up precipitated the single largest air battle of the Solomon Islands Campaign. Some 120 Japanese aircraft sortied against 104 Allied defenders in a dogfight over Savo Island, Tulagi and Cape Esperance. The Allies scored a remarkable one-sided victory, downing 49 Zeros and 32 dive-bombers, against the loss of just six aircraft. While the Japanese were still able to commit large forces to the battle, the victory on 16 June continued the pattern of an increasingly one-sided battle for air supremacy. From April to early June 1943 the ratio of the Allies' kills-to-losses averaged about 3:1; on 12 June the Allies scored a 5:1 victory and ten days later the win ratio jumped yet again to 13:1.[164]

Even a three to one win/loss ratio for the Allies demonstrated that

---

[162]Arbon, J. *The Bismarck Sea Ran Red.* Marceline, Missouri: Walsworth Press, 1979. See also McAulay, Lex. *Battle of the Bismarck Sea.* New York: St. Martin's Press, 1991.

[163]Birdsall, Steve. *Flying Buccaneers: The Illustrated Story of Kenney's Fifth Air Force.* New York: Doubleday, 1977.

[164]Frank, Richard. *Guadalcanal: The Definitive Account of the Landmark Battle.* New York: Random House, 1990.

they had developed a substantial advantage over the far more experienced Japanese. It spoke volumes about the advances made by Allied equipment as well as the quality of their pilots in the first half of 1943. Much has previously been written about the Japanese penchant for developing a small cadre of highly trained pilots as against Allied development of much larger numbers of less trained airmen. This utter nonsense is easily refuted by examining the kill ratios of the two combatants.

The Grumman F4F Wildcat was an American carrier-based fighter aircraft that began service with both the United States Navy and the British Royal Navy in 1940. In British service it was known as the Martlet. At the beginning of the Pacific War, the Wildcat was the only effective fighter available to the United States Navy and Marine Corps. Although initially losing out to the Brewster Buffalo, the United States quickly replaced the Buffalo with the Wildcat. With a top speed of 318 mph, the Wildcat was inferior to the faster, more maneuverable, and longer-ranged Mitsubishi A6M Zero. However, the Wildcat held its own partly because, with relatively heavy armor and self-sealing fuel tanks, the Grumman air-frame could survive far more damage than its lightweight, unarmored Japanese rival. The F4F's ruggedness was coupled with superior American tactics, such as the Thatch Weaver that obviated many of the Zero's reported strengths. Four U.S. Marine Corps Wildcats played a prominent role in the defense of Wake Island in December 1941. USN and USMC aircraft formed the fleet's primary air defense during the Battles of Coral Sea and Midway, and land-based Wildcats played a major role during the Guadalcanal Campaign of 1942–43. It was not until 1943 that more advanced naval fighters capable of taking on the Zero on more even terms, the Grumman F6F Hellcat and Vought F4U Corsair, reached the South Pacific theater. In the hands of an expert pilot using tactical advantage, the Wildcat could prove to be a difficult foe even against the formidable Zero. This American ingenuity in combination with a strong and durable design resulted in a claimed air combat kill-to-loss ratio of 5.9:1 in 1942 and 6.9:1 for the entire war.

The F6F Hellcat was the U.S. Navy's most successful airplane of World War II, operating almost exclusively against Japanese aircraft in the

Pacific theater of operation.[165] Designed to take on the Japanese Zero, it was remarkably successful. The Hellcat had a 19:1 kill ratio versus Japanese aircraft. It was the best piston fighter kill ratio of all times. The F6F operated primarily from aircraft carriers, although there were a few land-based units. The F6F had been deployed only during the final two years of the war. Still, F6Fs managed to shoot down over 5,000 Japanese aircraft during 1943 to 1945. Between the beginning of 1943 and the end of the war, total Japanese aircraft production was about 40,000 units. However, most Japanese aircraft operated in places like China, Manchuria, Indochina, East Indies etc. where U.S Navy carrier aircraft were not encountered or F6F pilots might have amassed even greater numbers.

Even the number 5,168 is somewhat incomplete. U. S. Naval Aviation compiled combat aviation statistics in 1947. It credited the F6F with destroying 5163 fighters and bomber types only. However, for the period 1 September 1944 through 15 August 1945 it registered a total of 3,518 credits to F6Fs, which included 2,278 single engine fighters, 36 land based single engine reconnaissance aircraft, 515 bombers and torpedo bombers, 89 float-planes, 530 twin engine combat type airplanes, 17 flying boats, 36 transport aircraft, and 17 training planes. If one adds these secondary aircraft the F6F total rises to 5,358 confirmed kills

The F4U Corsair was next best American airplane with a kill-to-loss ratio of approximately 12:1. Known to the Japanese as "Whispering Death" the Corsair racked up a remarkable record. The F4U did have its shortcomings in some areas, particularly regarding aircraft carrier operations where landings were difficult. It was the British who finally figured out that the Corsair could be more successfully landed on carriers if the pilot made an off-center approach, and the aircraft was used on U.S. carriers during the latter part of the war. Where the Corsair really shined though was as a ground attack aircraft.

For the entire war combined, all theaters of operation included, the P-51 Mustang was most successful with 5,954 kills. Its success was followed by the Hellcat with 5,168 kills. The third-ranking fighter was the P-38, with 3,785 enemy shot down in all theaters of operation combined.

[165]The Pacific Theater of Operations included the Aleutian Islands, Central Pacific, South Pacific and Southwest Pacific operating areas.

Corsair aircraft claimed 2140 enemy aircraft destroyed. In all in the Pacific theater of operations American aircraft shot down 12,666 enemy aircraft.[166] These numbers relate exclusively to air to air combat and do not include enemy aircraft destroyed on the ground or lost under non-combat conditions.

By October 1943 it had become clear that the air battle over the Solomon Islands was taking its toll on the Japanese Navy Air Force (JNAF). The gradual erosion of the Guadalcanal Campaign was putting increasing pressure, not so much on the availability of aircraft, but on the availability of trained pilots. It was systematic of the entire structure of the Imperial Japanese General Staff that war was planned as a short-term project with the emphasis on attack. The Japanese Navy, even more than the Imperial Japanese Army, was particularly unprepared for a war of attrition; their psychology, inherited from their great victory at the Battle of Tsushima, was to focus energy and resources on the winning of a single great all-out engagement rather than planning for a long war. An American intelligence report written in that month noted that Japanese pilots made glaring tactical mistakes, were unnecessarily exposing themselves to gunfire, frequently became separated and lost mutual support, and at times seemed to be completely bewildered by events transpiring before them. Army Intelligence noticed that both bomber and fighter pilots had ceased to display the aggressiveness which marked earlier combat. Bombers ceased to penetrate to their targets in the face of heavy fire, as they had formerly done; they jettisoned bombs, attacked outlying destroyers, gave up attempts on massed transports in the center of a formation. Fighters broke off their attacks on Allied heavy and medium bombers before getting within effective range, and often showed a marked distaste for close-in contest with Allied fighters.

Not only was the Japanese Navy failing to train sufficient pilots to man its ships at sea, it was also failing to produce enough pilots to fill the basic needs of its island possessions. Allied airmen rarely had to contend with Japanese warplanes when the Allies invaded a Japanese-held island. Only around Truk did American airmen experience much

---

[166] https://ww2aircraft.net/forum/threads/which-us-wwii-fighter-shot-down-the-most-enemy-aircraft.30227/

opposition from JNAF.

The JNAF was also failing to protect its pilots. It is instructive to consider that very few American pilots died as a result of their planes being shot down. In part that was because, unlike their Japanese counterparts, U. S. fighters had better armored cockpits. Moreover, without self-sealing fuel lines, Japanese Zero frequently blew up when hit by tracer bullets, killing the pilot instantly. When U. S. pilots ditched or parachuted into the sea, the U. S. Navy had a well-organized search and recovery capability. The Japanese Navy did not. Advantageously most of the dogfights in the April–June 1943 period took place closer to Allied held areas. American pilots were also better conditioned, with rotation and rest and recreation (R&R) built into the whole logistic framework of the various forces operating under COMAIRSOL.

Japanese fighters also had to face a multiplicity of challenges because of the diversity in capability of the six types of Allied fighter planes with which they had to engage. By contrast Allied pilots in the South Pacific only had to develop tactics to combat the Zero. At least two American aircraft were developed specifically to combat the Zero.[167]

The Japanese Navy

Japanese military planners first began plotting strategy against defined hypothetical enemies in 1907. Initially, those presumed enemies were Russia, the United States, and France. From the geo-strategic standpoint, the Army would have the major role in a war against Russia, while the Navy would be pitted against the United States and France. While all such plans allow for, even require, a few occasional revisions, this war plan remained nearly unchanged until 1936, when France was removed from the list of hypothetical enemies and China and Great Britain were substituted. It is indeed odd that the Dutch, masters of Indonesia, were not included. It is even more odd that Great Britain was ignored until mid-1930s. The hypothetical war with China would have been the responsibility of the Imperial Japanese Army, while that with Great Britain would have involved essentially a naval operation. Planning was

[167]https:// japaneseaircraftofwwii.blogspot.com / 2016 / 08 / degraded-performance-of-japanese-navy.html

predicated on the basic assumption was that Japan would be fighting only a single enemy, not two or three enemies simultaneously.

In the event of war with the United States, the plan, heavily based on Mahan's geopolitics, called for the Imperial Japanese Navy (IJN) to destroy the enemy's Far Eastern fleet at the outset of hostilities. Japanese naval units would then occupy Luzon and Guam in cooperation with the Army. Japanese planning assumed that the United States would then move most of its remaining naval forces to the relief of the Philippines. IJN would move to intercept and destroy the main enemy fleet when it sailed forth to Far Eastern waters. This assumption was pure Alfred Thayer Mahan, as interpreted and applied by Japanese strategists. IJN's absolute conclusion was that the main U.S. fleet would absolutely necessarily have to come to the Western Pacific sooner or later to challenge the Japanese aggression, in which case Japanese naval; and air forces would then destroy the American fleet once and for all. This strategy had worked splendidly for Japan's main when it sailed against, and utterly destroyed, the Russian main fleet in the Battle of Tsushima in 1905.

Japan built its navy to correspond to its long-range planning. As late as 1939 the Imperial Japanese Navy continued to plan, scheme, and build on the basis of the fully anticipated great all-out battle between traditional type warships. It assumed that decisive battles had to be fought principally by the big guns of the battleships, augmented and supplemented by cruiser and destroyer attacks. Air attacks from carrier-based aircraft did not figure prominently in IJN planning. Naval intelligence relegated aircraft principally to information gathering and surveillance. IJN had been conceived, built, constructed, armed, and trained accordingly.

Japanese naval planning had concluded that a strength equivalent to 70 percent of the total strength of the U.S. Navy was a prerequisite for victory over the United States. It made the assumption that 30 percent of the main U.S. fleet would be deployed elsewhere, such as the Atlantic Ocean, and that Japanese naval power would destroy some portion of the American fleet before it reached the waters of the Far Eastern. The Washington Naval Treaty of 1922 had limited the construction and maintenance of battleships. IJN's battleships were limited to 60 percent of the U.S. strength. It was for this strategic reason that the Japanese Navy had made strenuous efforts to build up its auxiliary strength. Following condemnation for its Asian aggression, and seething under restrictions on the size and composition of its navy under the Washington Naval Treaty, Japan withdrew from the League of Nations in 1934, and renounced all of the treaty obligations. As it was, as early as 1934, IJN planned to build two monster battleships, to be equipped with 18-inch guns.[168] In 1940, simultaneous efforts were made to strengthen air and submarine forces.[169]

The Japanese Navy had developed some very good submarines with excellent torpedoes and well-disciplined crews, but they did not know how to utilize them properly. Neither did they learn that technique from their close allies in Germany. IJN' submarine warfare failed for two principal reasons. First, IJN did not produce enough submarines. Many submarines could have been produced just from the cost of one super-battleship. Instead of wasting resources building battleships, Japan could have built more submarines and had those submarines attack allied transport and supply ships instead of warships. Second, IJN ordered the submarines to focus on the wrong targets. As spectacular as it is to sink a warship, it would actually have benefited Japan's war effort much more had they sunk tankers, supply vessels and transports.[170]

At the beginning of the Second World War, the Imperial Japanese Navy was arguably the most powerful ocean-going force in the world. Its naval aviation corps consisted of 10 aircraft carriers manned by 1500 highly trained aviators. With its superior aircraft, such as the Mitsubishi Zero, it was reportedly the most highly trained and proficient force of its kind. At the beginning of World War II, the Imperial Japanese Navy boasted that it had created the finest naval aviation corps in the world. Japanese aircraft were at least the equals of anything then flying in any air force. In some cases, such as with the legendary Zero fighter, Japan

---

[168]Jentschura, Hansgeorg; Jung, Dieter; and Mickel, Peter. *Warships of the Imperial Japanese Navy, 1869–1945*. Annapolis, Maryland: United States Naval Institute, 1977.

[169]Campbell, John. *Naval Weapons of World War II*. Annapolis, Maryland: Naval Institute Press, 1985.

[170]https://www.historyandheadlines.com/10-worst-japanese-blunders-world-war-ii/

claimed that its aircraft were substantially better than any then flying. Japanese aircrews were superbly trained before deployment. Many had been battle tested in the conflict in China during the late 1930s. When war erupted in the Pacific in December 1941, the high command viewed its navy and naval air force as superior to those of any of its opponents. Initially, Japanese aviators scored stunning victories during the first six months of the war, beginning with the attack on Pearl Harbor, and continuing through the sinking of the British men-of-war *Prince of Wales* and *Repulse*. IJN then conducted fearsome raids on northern Australia. IJN went on a rampage through the Indian Ocean, especially in April 1942. Only after it was checked at the Battle of the Coral Sea and defeated soundly at Midway was IJN reduced to equal terms with the Allies. Nonetheless, throughout the war, the IJN's air force remained a potent weapon. Eventually, Japanese equipment was eventually outclassed by newer American models, and relentless attrition began to take its toll on pilot quality. Even in defeat, though, the IJN air command refused to disappear, finally immolating itself in the form of the Kamikaze air corps.[171]

Nothing tells us more about the reasons for the defeat of Japan than the destruction of its once great and powerful navy. And one can see much of that failure in the choice the Imperial Japanese Navy made to construct two huge battleships instead of a fleet of aircraft carriers. Almost immediately the Imperial Japanese Navy undertook the design of the new *Yamato* class of heavy battleships. IJN laid down the keel of *Yamato* at the Kure Naval Arsenal, Hiroshima, on 4 November 1937, in a dockyard that had to be adapted to accommodate her enormous hull. IJN maintained great secrecy, to the extent of building a protective canopy over the *Yamato* during construction. IJN launched *Yamato* on 8 August 1940.

The *Yamato* and her sister ship, *Musashi*, were the largest and most heavily armed battleships ever constructed. Both displaced 72,800 tonnes under a full load and both were armed with nine 18.1 inch main guns, the largest guns ever mounted on a warship although her shells were not as heavy as those fired by the British 18-inch naval guns of World War I. Her secondary battery comprised twelve 155-millimeter guns mounted in four

---

[171] http://www.combinedfleet.com/ijnaf.htm

triple turrets; and twelve 127-millimeter guns in six twin mounts. In addition, Yamato carried twenty-four 25-millimeter anti-aircraft guns. When refitted in 1944 and 1945 for naval engagements in the South Pacific, the secondary battery configuration was changed to six 155 mm guns and twenty-four 127 mm guns, and the number of 25 mm anti-aircraft guns was increased to 162.[172]

The first ship of its class, she was named for Yamato Province. Her sister ship was the *Mushashi*, commissioned in mid-1942, Musashi was modified to serve as the flagship of the Combined Fleet, was also named after a Japanese province. *Musashi* had a length of 800 feet, 6 inches, between perpendiculars and 862 feet, 10 inches, overall. She had a beam of 121 feet, 1 inch and a draft of 35 feet, 8 inches, at deep load. *Mushashi* displaced 64,000 long tons. Her crew consisted of 2,500 officers and enlisted men in 1942, and about 2,800 in 1944. The battleship had four sets of Kampon geared steam turbines, each of which drove one propeller shaft. The turbines were designed to produce a total of 150,000 shaft horsepower using steam provided by 12 Kampon water-tube boilers, which gave her a maximum speed of 27 knots. She had a range of 7,200 nautical miles at the speed of 16 knots. *Musashi* was fitted with two catapults on her quarterdeck and could stow up to seven float-planes in her hangar which was located below-decks. The ship operated Mitsubishi F1-M biplanes and Aichi E13-A1 monoplanes and used a stern-mounted crane for aircraft recovery.[173]

In June 1942 *Yamato* served Admiral Isoruku Yamamoto[174] as his flagship during disastrous defeat at the Battle of Midway, although she played no actual role in the battle. She then moved to Kure naval base. *Yamato* left Kure for Truk on 17 August 1942. After 11 days at sea, she was sighted by the American submarine *USS Flying Fish*, which fired four torpedoes, all of which missed; Yamato arrived safely at Truk later that day. She remained there throughout the Guadalcanal Campaign because of a lack of appropriate ammunition for shore bombardment, and her

---

172Skulski, Janusz. *The Battleship Yamato: Anatomy of the Ship*. Chrysalis Books, 1988.

173Brown, David. *World War II Fact Files: Aircraft Carriers*. New York: Arco, 1977.

174Agawa, Hiroyuki. *The Reluctant Admiral: Yamamoto and the Imperial Navy*. Kondasha International, 1979.

enormous fuel consumption. The battleship spent only a single day away from Truk between her arrival in August 1942 and her departure on 8 May 1943. On that day, she set sail for Yokosuka and from there for Kure, arriving on 14 May. She then spent nine days in dry dock for general repairs, and in late July for significant refitting and upgrades. On 16 August, *Yamato* began her return to Truk, where she joined the defensive task force that IJN had formed in response to American raids on the Tarawa and Makin atolls.[175]

IJN assigned *Musashi* to the Combined Fleet, commanded by Admiral Isoroku Yamamoto, on 15 January 1943. She sailed for Truk three days later, arriving on 22 January. On 11 February, she replaced her sister ship *Yamato* as the fleet's flagship. On 3 April, Yamamoto left *Musashi* and flew to Rabaul, New Britain, to personally direct "Operation I-Go," the planned Japanese aerial offensive in the Solomon Islands. American intelligence intercepted Yamamoto's orders which Operation Magic deciphered. Upon direct orders from President Roosevelt in Operation Vengeance, American Lockheed P-38 Lightning fighters shot down Yamamoto's aircraft and killed him while he was en route from New Britain to Ballale, Bougainville. On 23 April, IJN moved his cremated remains back to Truk and placed them in his cabin on board *Musashi.*

On 17 May 1943, in response to American attacks on Attu Island, on the American Aleutian Island chain, Musashi, along with the carrier Hiyō, two heavy cruisers, and nine destroyers, moved to the northern Pacific. When they failed to make contact with the American naval forces, the ships sailed to Kure on 23 May, where Yamamoto's ashes were taken from the vessel in preparation for the formal state funeral. Next, Musashi's task force, now significantly reinforced for the counterattack on American naval forces off Attu, sailed again to confront the enemy. By this time the United States had captured the island. On 24 June, while being overhauled at Yokosuka Naval Arsenal, Musashi was visited by Emperor Hirohito along with several high-ranking naval officers. The ship was then equipped with a pair of Type 22 radars at Kure. She sailed for Truk on 30 July and arrived there six days later, where she resumed her position as fleet flagship for Admiral Mineichi Koga.

---

[175]Stille, Mark. *USN Carriers vs IJN Carriers.* Oxford, UK: Osprey, 2007.

In mid-October 1943, in response to suspicions of planned American raids on Wake Island, *Musashi* led a large fleet, consisting of three carriers, six battleships, and eleven cruisers, to intercept American forces. Having failed to make contact with the suspected American force, the *Musashi* task force returned to Truk on 26 October, where she spent the remainder of 1943 in Truk Lagoon. The deployment of such a large force on what was essentially a wild goose chase, severely cut into supplies of oil. Stockpiles of petroleum products were no longer easily replenished because of American air and submarine attacks on supply ships.

Meanwhile, *Yamato* spent the remainder of 1943, and much of 1944, moving between the major Japanese naval bases of Truk and Kure in response to American threats.

Subsequently, in late 1944, because of their extensive storage capacity and thick armor protection, IJN recruited *Yamato* and *Musashi* to serve as transport vessels. On 25 December, while ferrying troops and equipment, which were wanted as reinforcements for the garrisons at Kavieng and the Admiralty Islands, from Yokosuka to Truk, the U.S. submarine *Skate* intercepted *Yamato. Skate* fired a spread of four torpedoes at Yamato, one of which struck the battleship's starboard side toward the stern, ripping a huge hole that caused the rear turret's upper magazine to flood. Despite this, *Yamato* reached Truk later that day. IJN repair ship *Akashi* effected temporary repairs, and Yamato departed on 10 January for Kure. On 16 January 1944, *Yamato* arrived at Kure for permanent repairs to the torpedo damage. She was again dry-docked at Kure for further upgrades to all her radar and anti-aircraft systems from 25 February until 18 March 1944.

Yamato left the dry dock on 18 March 1944 and went through testing, beginning on 11 April. She left Kure on 21 April and embarked soldiers and materiel the following day at Okinoshima for a mission to Manila, reaching the Philippines on 28 April. She then moved on to Malaya to join Vice-Admiral Jisaburo Ozawa's Mobile Fleet at Lingga. This force arrived at Tawi Tawi on 14 May 1944. In early June, IJN again commandeered *Yamato* and *Musashi* as troop transports, this time to reinforce the garrison and naval defenses of the island of Biak as part of Operation Kon. However, IJN canceled that mission when Admiral Ozawa learned of American carrier attacks on the Mariana Islands. The Imperial Japanese Navy reorganized, on order to be able to concentrate its remaining fighting strength in the hope of achieving a decisive success against the Americans. From 19 to 23 June 1944, *Yamato* escorted forces of Admiral Ozawa's Mobile Fleet during the Battle of the Philippine Sea, known to American pilots as "The Great Marianas Turkey Shoot." At that engagement the Japanese lost three aircraft carriers and 426 aircraft. Although present at the Battle of the Philippine Sea in June 1944, *Yamato* again played no part in the battle.

As for *Musashi,* she remained in Truk Lagoon until 10 February 1944, when she returned to Yokosuka. On 24 February, *Musashi* sailed for Palau, carrying an Imperial Japanese Army battalion and also a battalion of Special Naval Landing Forces along with all their equipment. After losing most of her deck cargo during a typhoon, *Musashi* arrived at Palau on 29 February and remained there for the next month. On 29 March, Musashi departed Palau under cover of darkness to avoid an expected air raid. While escaping she encountered the submarine *USS Tunny,* which fired six torpedoes at the battleship; five of which missed. However, the sixth torpedo blew a hole almost twenty feet in diameter near the bow. Following temporary repairs, *Musashi* sailed for Japan and arrived at Kure Naval Arsenal on 3 April. In May 1944, *Musashi* departed Kure for Okinawa on 10 May, then for Tawi-Tawi on 12 May. She was assigned to the 1st Mobile Fleet. On 10 June, the battleships departed Tawi-Tawi for Batjan in preparation for Operation Kon, a planned counterattack against the American invasion of Biak. Two days later, when word reached IJN of American attacks on Saipan, IJN diverted this force to the Mariana Islands. During the Battle of the Philippine Sea, Musashi served no martial function but neither was she attacked.

Following the Battle of the Philippine Sea, Yamato withdrew with the Mobile Fleet to the Hashirajima staging area near Kure to refuel and rearm. With *Musashi* she left the fleet on 24 June for the short journey to Kure, where she received five more triple 25 mm anti-aircraft mounts. IJN used this opportunity to order "emergency buoyancy keeping procedures," which resulted in the removal of almost every flammable item from the battleship, including linoleum, bedding, and mattresses. As a result, men slept on planks which could be used to repair damage. Flammable paints received a new silicon-based overcoat, and IJN installed additional portable pumps and firefighting apparatus.

Between 22 and 25 October 1944, as part of Admiral Takeo Kurita's Center Force, which was also known as Force A or First Striking Force, Yamato took part in one of the largest naval engagements in history, known as the Battle of Leyte Gulf. In response to the American invasion of the Philippines, Operation Shō-Gō called for a number of Japanese groups to converge on the island of Leyte, where American troops were landing. On 18 October, IJN gave *Yamato* a coating of black camouflage in preparation for her nighttime transit of the San Bernardino Strait. The principal ingredient of this blackening was soot taken from her smokestack. While en route to Leyte, the American submarines *USS Darter* and *USS Dace* attacked the task force in the Palawan Passage on 23 October 1944. The submarines sank two Takao-class heavy cruisers including Kurita's flagship, *Atago*, and damaged a third. Admiral Kurita survived the loss of *Atago* and transferred his flag to *Yamato*.

The following day at the Battle of the Sibuyan Sea American forces severely damaged the Center Force. It lost three more heavy cruisers, eliminating a substantial part of the fleet's anti-aircraft defense. During the course of the day, American carrier aircraft sortied a total of 259 times. Aircraft from the American carrier *USS Essex* struck *Yamato* with two armor-piercing bombs and scored one near miss. *Yamato* suffered moderate damage, but remained battle-worthy. The only time *Yamato* fired her main guns at enemy surface targets was in October 1944, when she was sent to engage American forces invading the Philippines during the Battle of Leyte Gulf. On the verge of success, the Japanese force turned back, believing they were engaging an entire U.S. carrier fleet rather than a light escort carrier group that was all which stood between the battleship and vulnerable troop transports. *Yamato* thus emerged from the battle without serious damage; only three near misses from bombs and seventeen casualties from strafing were suffered during the battle itself, while carrier strikes during the retreat caused light damage to the ship and injured or killed 21 crewmen. Three more heavy cruisers and one light cruiser were subsequently lost.

However, *Musashi* became the focus of the American air attacks and eventually sank. There is no question that the Yamato class battleships were well constructed. During the attack on *Musashi*, American aircraft hit the ship with an estimated total of 19 torpedoes and 17 bombs. *Musashi*'s commander, Admiral Inoguchi, chose to go down with his ship. Other ships recused 1,376 of her 2,399-man crew, about half of which survivors were evacuated to Japan, while the rest took part in the Japanese defense of the Philippines. The destroyer *Shimakaze* rescued survivors from the *Maya* which Americans had sunk earlier, and who had received temporary haven aboard the *Musashi.*

On 15 November 1944, the 1st Battleship Division was disbanded and *Yamato* became the flagship of the Second Fleet. On 21 November, while transiting the East China Sea in a withdrawal to Kure Naval Base, the submarine *USS Sea Lion* attacked the *Yamato*'s task force. Equipped with the new and now highly effective torpedoes the Sea Lion sunk the the battleship *Kongō* and destroyer *Urakaze. Yamato* was immediately dry-docked for repairs and anti-aircraft upgrades.

On 3 January 1945, *Yamato* left dry-dock two days later for Japan's Inland Sea. IJN deactivated the first Battleship Division again on 10 February and reassigned *Yamato* to the 1st Carrier Division. On 19 March, American carrier aircraft from the carriers *Enterprise, Yorktown* and *Intrepid* raided Kure. Although the American aircraft scored hits on sixteen Japanese warships, *Yamato* sustained only minor damage.

In April 1945, Yamato was dispatched on a one-way mission to Okinawa in April 1945, with orders to beach herself and fight until destroyed protecting the island. Operation Ten-Go was the final commitment of Japan's remaining naval surface strength. *Yamato* and nine escorts, including the cruiser Yamaha along with eight destroyers, set sail for Okinawa. This small fleet was ordered to act in concert with kamikaze and Okinawa-based army units, and attack the Allied forces assembled on and around Okinawa. *Yamato* would then be beached to act as an unsinkable gun emplacement and continue to fight until destroyed. American submarines and air patrols identified the task force south of Kyushu by US submarines and aircraft, and moved to intercept. On 7 April 1945 she was sunk by American carrier-based bombers and torpedo bombers with the loss of most of her crew.[176]

Japan had constructed the twin giants of the sea, *Yamato* and *Mushashi,* at considerable cost. Unlike the United States, with its huge resources, Japan could ill afford to waste its precious resources in the construction of useless or superfluous ships. Their construction was the result of backward and dated thinking among the IJN policy makers. Already the aircraft carrier, not the battleship, was the seat of sea power. Forward thinking might have allowed the construction of several large and effective carriers, along with other modern and far more useful weapons systems. As it was, neither super-battleship had even the slightest impact on the conduct of the Pacific War.

Initially IJN had planned to build three *Yamato* class super-battleships. Construction of the *Shinano* as a battleship began in May 1940. However, following Japan's disastrous loss of four fleet carriers at the Battle of Midway in mid-1942, IJN altered *Shinano's* partially finished hull so it was to be completed as a carrier. The conversion was still not finished in November 1944 when she was ordered to sail from the Yokosuka Naval Arsenal to Kure Naval Base to complete fitting out and transfer a load of 50 Yokosuka MXY7 Ohka rocket-propelled kamikaze flying bombs. However, 10 days after commissioning, on 29 November 1944, the U.S. Navy submarine *Archerfish* intercepted the *Shinano* and sank it. Over a thousand sailors and civilians were rescued but 1,435 went down with the ship, including her captain. *Shinano* was the largest warship ever sunk by a submarine.[177]

---

[176] The accounts of the two super-size Japanese battleships generally follows these works: Garzke, William H. and Dulin, Robert O. *Battleships: Axis and Neutral Battleships in World War II.* Annapolis, Maryland: Naval Institute Press, 1985; Jackson, Robert. *The World's Great Battleships*. Brown Books, 2000; Jentschura, Hansgeorg; Jung, Dieter; and Mickel, Peter. W*arships of the Imperial Japanese Navy, 1869–1945.* Annapolis, Maryland: United States Naval Institute, 1977; and Johnston, Ian and McAuley, Rob. *The Battleships.* MBI Publishing Company, 2000.

[177] Enright, Joseph F. & Ryan, James W. (1987). Shinano!: The Sinking of Japan's Secret Supership. New York: St. Martin's Press.

Units 731 and 100

Of all the belligerents in World War II, Japan was the only nation that used both chemical and biological weapons in battle. The other major combatants possessed stores of chemical weapons, but did not commit any on the battle field. Some countries either possessed the means for producing biological weapons, or had crash programs seeking to develop such unconventional tools of war. However, none were employed against the enemy before Japan surrendered in mid-August 1945. Japan, under the growing influence of ultra-nationalist militarists, began research on both chemical and biological weapons in the late 1920s. Although many of the ultra-nationalists who were in a position of influence within the IJA endorsed investigation of the possibilities of utilizing germ and chemical warfare in future conflicts, the man most responsible for launching Japan's chemical and biological weapons programs was a young Army doctor, Shiro Ishii. A charismatic, charming, unscrupulous, amoral, but brilliant researcher, Ishii held both medical and Ph.D. degrees from Kyoto Imperial University, one of Japan's premier institutions of higher learning. Ishii achieved a meteoric rise in the Army, gaining promotion to higher rank every three years during his military career. At the time of Japan's surrender, he held the rank of Lieutenant-general, the highest rank awarded in the Medical Corps. From the beginning of his military career, Ishii enjoyed the patronage of key figures within the Japanese High Command. As such, he avoided virtually any supervision of his activities, and was free of any restraint in conducting the most sordid investigations in pursuit of the development of chemical and biological weapons. From the start, Ishii engaged in involuntary human experiments.

Unit 731 was a covert biological and chemical warfare research and development unit of the Imperial Japanese Army that undertook lethal human experimentation during thc Second Sino-Japanese War (1937–1945) part of World War II. It was responsible for some of the most notorious war crimes carried out by Imperial Japan. Unit 731 was based at the Pingfang district of Harbin, the largest city in the Japanese puppet state of Manchukuo (now Northeast China). It was officially known as the Epidemic Prevention and Water Purification Department of the Kwantung Army.[178]

Experimentation initially was implemented in the Zhongma Fortress, a prison and experimentation camp in Beiyinhe, a village 62 miles south of Harbin located conveniently on the South Manchuria Railway. A jailbreak in autumn 1934 and later explosion, believed to have been a Chinese attack, in 1935 led Ishii to close down Zhongma Fortress. The Imperial Japanese Army granted the authorization to move to Pingfang, approximately 15 miles south of Harbin, and build a much larger facility. In 1936, Emperor Hirohito, by decree, personally authorized the expansion. Simultaneously, it was divided into the Ishii Unit and also the Wakamatsu Unit with a base in Hsinking. Originally set up under the Kempeitai military police of the Empire of Japan, Unit 731 was taken over and commanded until the end of the war by General Shirō Ishii, a combat medic officer in the Kwantung Army. The facility itself was built between 1934 and 1939 and officially adopted the name "Unit 731" in 1941.[179]

IJA gave Ishii enormous resources in men, material, and money to conduct his work in Manchuria. It also gave Ishii permission to use various subterfuges in order to conceal secretly his real activities. Ishii originally named his research unit, the Togo Unit, after one of his heroes of the Russo-Japanese War. Later, the unit's name was changed to the Ishii Unit. Another alias was the Water Purification Unit. It is known today simply as the infamous Unit 731. Once research in laboratories showed promising results, Ishii, his superiors, and his collaborators agreed that more extensive tests were needed in order to prove the viability of bacterial and chemical weapons being developed. Consequently, field tests were

---

[178]Harris, Sheldon H. "Japanese Biomedical Experimentation during the World War II Era" *Military Medical Ethics* 2: 463-506. See also Tanaka Y. *Hidden Horrors: Japanese War Crimes in World War II*. Boulder, CO: Westview Press; 1996; and Tsuneishi K. *The Germ Warfare Unit That Disappeared: Kwantung Army's 731st Unit*. Tokyo: Kai-mei-sha Publishers; 1981.

[179]"Unit 731" entry in Wikipedia. Also "Unit 731 - Did the Emperor Know?" British Television South documentary first broadcast on 13 August 1985.

ordered beginning in 1939, if not earlier. Bacteriological and chemical artillery shells, bombs, canisters and barrels filled with deadly pathogens were used in border skirmishes with the Soviet Union in the summer of 1939. Later, from 1940-1942, extensive field tests were conducted throughout Manchuria (Manchukuo, as the puppet state was known), and in Eastern and Central China. Many cities and towns were saturated with pathogens disseminated by aerosol spraying from airplanes, by lacing water wells with germs, and by providing innocents with food, scraps of cloth, and sweepings of wheat, rice and assorted fruits and vegetables that were filled with a host of different germs.

Japanese Unit 731 has remained one the most closely guarded secrets of Japanese atrocities of World War II. The Japanese government did not acknowledge the existence of Unit 731 until recently. Finally, in 1984 Japan acknowledged that its military had indeed executed the most vile experiments on human beings that could possibly be imagined. The staff of Unit 31 deliberately infected human beings with bubonic plague, anthrax, cholera and other pathogens. Because of the Unit's secret nature, there is no complete list of the experiments that were undertaken by Unit 731. Physicians and researchers performed invasive surgery on prisoners, removing organs to study the effects of disease on the human body. These were conducted while the patients were alive because it was thought that the death of the subject would affect the results.[180]

The best estimate is that at least 3,000 soldiers and civilians were used as guinea pigs. Some of the more horrific experiments included vivisection without anesthesia and pressure chambers designed to see how much a human could take before his eyes popped out. Japanese authorities, the unit used thousands of Chinese and other Asian civilians and wartime prisoners as human guinea pigs to breed and develop killer diseases. Another medical experiment employed barbaric trials to determine the effect and possible treatment of frostbite on the human body.[181]

---

[180] See also Tsuneishi K, Asano T. *The Bacteriological Warfare Unit and the Suicide of Two Physicians*. Tokyo: Shincho-Sha Publishing Co; 1982. English translation located in the Fort Detrick Archives, Fredrick, Maryland.

[181] Roland, C. G. "Human Visection: The Intoxication of Limitless Power in Wartime" in Moore, B, and Fedorowich, K, eds. *Prisoners of War and Their Captors in World War II.* Oxford: Berg; 1996.

The Japanese created Unit 731 in 1938 in Japanese-occupied China with the aim of developing biological weapons. It also operated a secret research and experimental school in Shinjuku, central Tokyo and was headed was Lieutenant Shiro Ishii [later general]. Various Japanese universities and medical schools supplied the technicians, physicians, surgeons, and research staff.[182]

> Spurred on by Lieutenant-General Ishii Shiro, 3,000 Japanese researchers working at Unit 731's headquarters in Harbin infected live human beings with diseases such as the plague and anthrax and then eviscerated them without anesthesia to see how the diseases infected human organs. They frequently eviscerated the subjects without anesthesia to see how the diseases infected human organs. The Japanese would commonly test new biological weapons such as plague cultures or bombs filled with plague-infested fleas on them.[183]

In 1984, a graduate student at Keio Medical University in Tokyo found records of human experiments in a bookstore. The pages described the effects of massive dosages of tetanus vaccine. There were tables describing the length of time it took victims to die and recorded the muscle spasms in their bodies.[184]

Presumably in order to ease the conscience of those involved, the prisoners were referred to not as people or patients but as *Maruta* [wooden logs]. At least some 3,000 people, not just Chinese but also Russians, Mongolians and Koreans, and a few Americans and Australians and British POWs died from the experiments performed by Unit 731 between 1939 and 1945. No prisoner came out alive of the Unit's gates.

Since the Japanese army used poison gas during the war in China, one of the charges to Unit 731 was to develop a more potent poison gas. The way to test the efficiency and effectiveness was to use the prisoners in Unit 731 as test subjects.

---

[182]Powell J. W., Jr. "Japan's Biological Weapons, 1932–1945." *Bulletin of Atomic Scientists*. October 1981.

[183]Handelman, Stephen and Alibek, Ken. *Bio-hazard: The Chilling True Story of the Largest Covert Biological Weapons Program in the World—Told from Inside by the Man Who Ran It*, Random House, 1999.

[184]Barenblatt, Daniel. *A Plague Upon Humanity: The Secret Genocide of Axis Japan's Warfare Operation.* Harper Collins, 2005.

During the war, the Japanese Imperial Army used biological weapons developed and manufactured by Unit 731's laboratory in Harbin throughout China, killing or injuring an estimated 300,000 people. Unit 731 and also Unit 100 were the two biological warfare research centers set up in clear violation of the Geneva Protocol of 1925 banning biological and chemical warfare.

Japanese physicians forced sex acts between infected and non-infected prisoners with an eye to creating a method to prevent transmission of syphilis between patients. A male and female, one infected with syphilis, would be brought together in a cell and forced into sex with each other. Anyone resisting would be shot. After victims were infected, they were vivisected at different stages of infection, so that internal and external organs could be observed as the disease progressed. Genitals of female prisoners that were infected with syphilis were called "jam filled buns" by guards. Some of the children who grew up inside Unit 731 were also infected with syphilis.

In other tests, subjects were deprived of food and water to determine the length of time until death; placed into high-pressure chambers until death; experimented upon to determine the relationship between temperature, burns, and human survival; placed into centrifuges and spun until death; injected with animal blood; exposed to lethal doses of x-rays; subjected to various chemical weapons inside gas chambers; injected with sea water; and burned or buried alive. There was no thought of humane treatment of the victims.

Long after the war was over one former minor technician explained why he participated in the gruesome experiments at Unit 731. "Before the war, we were told we'd have to fight to protect ourselves, our families and the Japanese people. These people recount their vivid firsthand memories of what it was like to cut open pregnant women as they lay awake on the vivisection table, inject plague germs into healthy farmers, and carry buckets of fresh blood and organs through corridors to their appropriate destinations." After four years of training with, and brain-washing by, the unit, he knew what was happening at Harbin and was not surprised at what he saw in the cell. Still, he had not even felt guilt while being an active part of the horror.[185]

When it became apparent that Japan would be defeated, Japanese soldiers destroyed the headquarters of Unit 731. Before Japan's surrender, the Imperial Japanese Army destroyed the site of the experiments so that no evidence was left. The Japanese Army executed the remaining 400 prisoners so that they could not bear witness to the so-called experiments. Japanese civilian and military employees of the unit had to swear secrecy. The Army also released the disease-infected mice the staff kept in the laboratory. Reportedly, the mice spread disease that cost the lives of at least 30,000 Chinese people, nearly all civilians.

Few of those involved with Unit 731 ever admitted their guilt. Some lower level staff and technicians were caught in China at the end of the war. Some were arrested and detained, but only a handful of them were prosecuted for war crimes, and those were tried by the Soviet Union, not the Western Allies.[186]

A few of the unlucky Unit 731 officers and enlisted men were arrested by Soviet forces and tried at the Khabarovsk War Crime Trials. The Japanese were also put on trial by the Soviet Union and China. The Soviet Union conducted a tribunal of twenty Japanese servicemen in the city of Khabarovsk in late 1949 and found twelve of them guilty of waging bacterial warfare. Most of those defendants were executed. Some of those in Unit 731 had committed suicide as all Japanese personnel were issued cyanide capsules as the Russians approached. Apparently, the Chinese Nationalist government tried and executed no one for the events at this facility despite the fact that most human subjects were Chinese. In the mid-1950s, Communist China tried several hundred Japanese officers and civilians, many of whom had been transferred from the Soviet Union. Most of these later defendants were given reduced sentences in return for admission of guilt, and none received the death penalty. [187]

---

[185] Gold, Hal. *Unit 731 Testimony: Japan's Wartime Human Experimentation Program.* Tuttle, 2004.

[186] *Khabarovsk War Crime Trials. Materials on the Trial of Former Servicemen of the Japanese Army Charged with Manufacturing and Employing Biological Weapons*, Moscow: Foreign Languages Publishing House, 1950.

[187] *Materials on the Trial of Former Servicemen of the Japanese Army, Charged with Manufacturing and Employing Bacteriological Weapons.* Moscow: Foreign

In 1946, U. S. General Douglas MacArthur granted all the Japanese scientists and physicians from Unit 731 immunity from war crimes prosecution in exchange for the germ warfare data gathered from experiments in Harbin. As explained in an internal War Department memorandum, dated June 23, 1947: "Since it is believed that the USSR possesses only a small portion of the technical information, and since any war-crimes action would completely reveal such data to all nations, it is felt that such publicity must be avoided in the interests of defense and security of the U. S. . . . It is believed also that the war-crimes prosecution of Gen. Ishii and his associates would serve to stop the flow of much additional information of a technical and scientific nature."[188]

None of the Japanese scientists in Unit 731 was ever punished. General Ishii lived in relative comfort and with respect, on the outskirts of Tokyo until his death in 1959. Other successful former operatives of Unit 731 include the former governor of Tokyo, the former president of the Japan Medical Association, the former director of the health ministry's preventive health research center, the former chairman and president of Green Cross Corp. and the past heads of a number of Japanese medical schools. The man in charge of vivisection, Yoshisuke Murata, became director of the respected Kyoto University medical school, and later medical director at Kinki University.[189]

In Japan, not one was brought to justice. In a secret deal, the post-war American administration gave them immunity for prosecution in return for details of their experiments. Some of the worst criminals, including Hisato Yoshimura, who was in charge of the frostbite experiments, went on to occupy key medical and other posts in public and private sectors in the home islands after the war.

Shirō Ishii (1892-1959) was a Japanese medical officer who held the rank of Surgeon General. By training he was a microbiologist. General Ishii made extensive visits to Europe in order to tour military hospitals during the mid-1920s, but the main reason for his European tour was to

---

Languages Publishing House, 1950.

[188] Harris, Sheldon H. *Factories of Death: Japanese Biological Warfare, 1932-45 and the American Cover-Up.* Routledge, 2004.

[189] Richard Lloyd Parry. "Dissect them alive: order not to be disobeyed." *London Times*, 25 February 2007.

investigate biological warfare research. Ishii studied the findings that were published during World War I. This research later heavily influenced his experiments in Unit 731. Additionally, on his world tour, some of the countries he visited were secretly researching biological warfare. Ishii organized a secret research group, the *Tōgō Unit*, to conduct various chemical and biological weapons. Ishii had proposed the creation of a Japanese biological and chemical research unit in 1930, after a two-year study trip abroad, rationalizing it on the grounds that Western powers were developing their own programs.

One of Ishii′s main supporters inside the army was Colonel Chikahiko Koizumi, who became Japan's Health Minister, serving from 1941 until 1945. Koizumi had witnessed and had been impressed by the successful German use of chlorine gas at the Second Battle of Ypres. A Military Attaché in the Japanese Embassy in Washington, D.C., later reported that Ishii had studied bacteriological warfare at the Massachusetts Institute of Technology in Boston. Upon his return to Japan in 1930, he convinced the Japanese State Department of Military Affairs to develop a biological weapons program. He became the director of Unit 731, heading the biological warfare unit of the Imperial Japanese Army involved in forced and frequently lethal human experimentation. Like many other former scientists at Unit 731, he was granted immunity and recruited by the United States and continued to conduct his research after the Second World War ended. Ishii was promoted to senior army surgeon, first class, surgeon colonel, in 1938. On 9 February 1939, he gave a lecture on bacteriological warfare in the War Ministry Grand Conference Hall in Tokyo. One of those attending was Prince Yasuhito Chichibu, a brother of Hirohito, the Emperor of Japan. Chicibu watched vivisection demonstrations by Ishii. For his services the Emperor endowed Ishii with the Order of the Golden Kite. Occupation authorities arrested Ishii at the end of World War II.[190]

The Soviet Union intended to interrogate and then try for war crimes and crimes against humanity both Ishii and the other Unit 731 leaders. Instead, Ishii and his team managed to escape from Manchukuo

---

[190]Tsuchiya, Takashi, "The Imperial Japanese Experiments in China" in *Oxford Textbook of Clinical Research Ethics*, 2008.

and to negotiate full immunity in 1946 from war-crimes prosecution. They had offered in exchange their full disclosure of germ warfare data based on human experimentation. Although the Soviet authorities wished the prosecutions to take place, the United States General Douglas MacArthur, after reviewing the reports of the investigating US microbiologists, found that the materials Ishii offered were so valuable as to grant immunity. Among these was Dr. Edwin Hill, Chief of Fort Detrick, Maryland, whose report stated that the information was "absolutely invaluable" and that it "could never have been obtained in the United States because of scruples attached to experiments on humans," and "the information was obtained fairly cheaply." On 6 May 1947, MacArthur wrote to Washington that "additional data, possibly some statements from Ishii probably can be obtained by informing Japanese involved that information will be retained in intelligence channels and will not be employed as 'War Crimes' evidence." The deal was not concluded until 1948. Richard Drayton, a Cambridge University history lecturer, claimed that Ishii later went to Fort Detrick, Maryland, to advise American scientists on bio-weapons.[191]

Japan had established its chemical weapons manufactory in Okunoshima, Hiroshima Prefecture in the early 1930s. Experiments by Ishii's team led to the creation of even more deadly agents that could then be manufactured at Okunoshima. Reportedly, Japan used chemical warfare in 1939 in Shanshi Province, China, against the Soviet Red Army. Nanking became the center of Japanese research and development in chemical and bacteriological warfare.[192] In what was originally a Chinese hospital the Japanese bred disease-bearing fleas and lice. And there were sporadic reports of the use of germ warfare and also chemicals against the Chinese in the early 1940s, especially in Chekiang Province. Both American President Franklin Roosevelt and British Prime Minister Winston Churchill then publicly threatened Japan with massive retaliation in kind if such weapons were used against Allied soldiers.[193] It is unclear if those threats included continued use against the Chinese Army and civilian

[191] "Ishi" entry in Wikipedia.

[192] Harris R, Paxton J. *A Higher Form of Killing: The Secret Story of Chemical and Biological Warfare.* New York: Hill and Wang; 1982.

[193] Allen and Polmar, *op. cit.,* 189-91.

population. General MacArthur suppressed knowledge of the use of germ, biological, chemical, and gas warfare in China. Since only Russian forces had visited and then occupied the infamous Unit 731 that gruesome experimentation could be suppressed also. MacArthur's staff was most successful in withholding information on any experimentation on Westerners at Unit 731.[194]

In 1981, journalist Seiichi Morimura published a highly fictionalized book entitled *The Devil's Gluttony*, which was based in part on interviews with several surviving Japanese who worked at the Unit 731 laboratory. It related several accounts of victims who had been subjected to such experiments as freezing, high pressure and vivisection. Morimura followed *The Devil's Gluttony* with *The Devil's Gluttony: A Sequel* in 1983. While these books purported to reveal the true operations of Unit 731, they actually confused these atrocities with those perpetrated in Unit 100. Too, the books used unrelated photos attributing them to Unit 731.[195]

In 1936 the Imperial Japanese Army created the Anti-Epizootic Protection of Horses Unit was created in Changchun, which became known as Unit 100. Unit 100 was an Imperial Japanese Army facility called the Kwantung Army Warhorse Disease Prevention Shop that focused on the development of biological weapons during World War II. It was operated by the Kempeitai, the Japanese military police. Major Yujiro, a veterinarian, commanded the facility. Yujiro occasionally cooperated with General Ishii of Unite 731 in various research projects. It was located in Mokotan, Manchukuo, a village just south of the city of Chagchun.

This facility, Unit 100, specialized in the prevention of animal diseases, but also researched plant and animal biological warfare and was thus akin to some of the activities of Unit 731. The unit focused on sabotage operations, with specific research conducted on the cultivation of crop viruses and livestock diseases. Yujiro's staff used both people and animals as subjects for experimentation. There exist a number of

[194] Allen and Polmar, *op. cit.*, 173-74.

[195] Nozaki, Yoshiko, *Textbook Controversy and the Production of Public Truth: Japanese Education, Nationalism, and Saburo Ienaga's Court Challenges. University* of Wisconsin—Madison, 2000, pp. 300, 381.

testimonies from former workers and medical students from Unit 100 which describe seeing dead bodies being wheeled away from operating rooms. Many animals were poisoned with contaminated crops.[196]

Theoretically, the main purpose of Unit 100 was to conduct research about diseases originating from animals. Since the Imperial Japanese Army was still heavily dependent on horses, the Imperial Japanese Army hoped to find ways to protect them. Simultaneously they wished to kill the enemy's horses and thereby weaken its military power. Furthermore, they hoped to spread disease via animal carriers. To this end, former members claim that experiments were also conducted with human beings. In practice, Unit 731 was the group tasked with developing biological weapons against humans.

Although smaller than Unit 731, Unit 100 was still a large organization. Its annual bacteria production capacity was projected to reach 1,000 kilograms of anthrax, 500 kilos of glanders, and 100 kilos of red rust fungus. These goals were never achieved, largely because of equipment and supply shortages.

Unit chief Yujiro Wakamatsu ordered his staff to purchase hundreds of cattle and put them to pasture along the Soviet border north-east of Hailar, ready to be infected by airborne dispersion. It was hoped that in the event of a Soviet invasion these infected livestock would mingle with local herds to cause epidemics and to destroy food supplies.[197]

In Soviet-sponsored war trials, former Unit 100 workers confirmed that they had "taken part in these inhuman experiments on living people, in bacteriological sabotage and in the preparations for bacteriological warfare against the Soviet Union." The main victims, however, were to be the Chinese. There were simply too many Chinese for the Imperial Japanese Army to handle. One way to depopulate an area was through chemical and/or biological warfare. One application of biological

---

[196] Cunliffe, William H. "Nazi War Crimes and Japanese Imperial Government Records: Select Documents on Japanese War Crimes and Japanese Biological Warfare, 1934-2006." Interagency Working Group.

[197] Drea, Edward; Bradsher, Greg; Hanyok, Robert; Lide, James; Petersen, Michael, & Daqing Yang. *Researching Japanese War Crimes: Introductory Essays.* Washington, D.C.: Nazi War Crimes and Japanese Imperial Government Records Interagency Working Group, 2006.

warfare was destruction of crops and the infiltration of crops with poison. Other admissions of guilt were found throughout the trial transcript.[198]

Some 180 Chinese victims and families of victims sued the Japanese government, demanding an apology, victim compensation and a full disclosure of the experiments in Unit 731. They failed in their suit as the High Court in Tokyo did finally acknowledge the existence of Unit 731 but upheld a lower court ruling that denied the plaintiffs compensation. And, of course, there was no apology.[199]

Unit 731 was not the only biological and chemical warfare unit that the IJA commissioned to aid Japan's military efforts. Dozens of Units similar to Ishii's were established under High Command orders, or, on several occasions, by Emperor Hirohito's acquiescence. The Units were to be found in virtually every area Japan conquered in the period 1931-1945. Some of the major units are known while others remain to be discovered. Among known chemical and biological warfare units were those located in Changchun, known as Unit 100, which specialized in plant and animal biological warfare; Beijing, designated Unit 1855, whose specialization was in in creating cholera epidemics; Hailar in Inner Mongolia, designated as Unit 2646, but also known as Unit 80, which engaged in various human subject experiments; and in Jilin Province in Manchuria we find Unit 516, whose specialty was chemical warfare, and which operated in conjunction with Ishii's Unit 731. Other units whose exact function and specialty are as yet unknown include Canton Guangzhou, called Unit 8604, which also operated also in Hong Kong; Nanking/Nanjing which hosted Unit Ei-1644; Singapore which had Unit 9420; and Shanghai, unit number unknown. Human experiments took place in Shanghai under the command of Lt. General Kitano Masaji, one of the most notorious advocates of bacterial warfare. Each of these units operated with between five and twelve support units. These support units were housed in small cities and villages throughout occupied territories. Among such places were Dairen (Dalian), Anda, and Sunyu. Each of these

---

[198] Yamada, Otozo. *Materials on the Trial of Former Servicemen of the Japanese Army, Charged with Manufacturing and Employing Bacteriological Weapons.* Moscow: Foreign Languages Pub. House, 1950.

[199] Felton, Mark. *The Devil's Doctors: Japanese Human Experiments on Allied Prisoners of War.* Pen & Sword, 2012.

smaller units was used to manufacture quantities of pathogens for use in the experimental laboratories and in the field testing. Other facilities manufactured the laboratory equipment used in conducting these gruesome tests. Still other facilities produced bomb casings, shell casings and canisters which Ishii's men could fill with either pathogens or chemicals at the larger facilities. Developing and implementing biological and chemical weapons required the IJA to allocate tremendous expenditure of men, material and money. This was done at a time when the government had but limited resources to expend.[200] Sheldon Harris, in his authoritative history of the Japanese biological warfare program, wrote concerning the Japanese response to the exposure of Isgii and his illegal experiments:

> The questions of ethics and morality as they affected scientists in Japan and in the United States never once entered into a single discussion... In all the considerable documentation that has survived..., not one individual is chronicled as having said [biological warfare] human experiments were an abomination and that their perpetrators should be prosecuted. The only concern voiced was that of the possibility of exposure that would cause the United States some embarrassment should word of the bargain ever become public knowledge.[201]

---

[200] Harris, Sheldon H. *Factories of Death: Japanese Biological Warfare, 1932-45 and the American Cover-Up.* Routledge, 2004. See also, generally, Williams, P. and Wallace, D. *Unit 731, The Japanese Army's Secret of Secrets*. London: Hodder & Stoughton; 1989.

[201] Harris, *Factories of Death*, p. 305.

Maltreatment of Asians

Japanese, resident in various Western Powers, had long suffered from racial discrimination issues. In 1919, Japan proposed a racial equality clause be added to the treaties ending World War I at the Paris Peace Conference. The majority of the victorious powers accepted Japan's proposal but American President Thomas Woodrow Wilson vetoed the idea, in violation of the rules of the Conference which upheld majority vote. Wilson's bigotry was followed in 1924, when the U.S. Congress enacted the Asian Exclusion Act, outlawing immigration from Asia. Little was heard of racial equality in Japan thereafter since it definitely did not fit with the emerging statism. Hirohito' reign became associated with the rediscovery of Hakkō ichiu as an expansionist element of Japanese nationalistic beliefs, and thus tied to Japanese racism.[202]

Racial discrimination against other Asians was habitual in Imperial Japan, having begun simultaneously with the rise of Japanese colonialism. With roots in its past, Meiji era Japan displayed a contempt for other Asians. Japan expected to treat other Asians exactly as the Western Powers treated them. The Shōwa regime preached racial superiority and racist theories.[203] According to historian Kurakichi Shiratori, one of Emperor Hirohito's teachers: "Therefore nothing in the world compares to the divine nature (shinsei) of the imperial house and likewise the majesty of our national polity (kokutai). Here is one great

[202] Bix, Herbert. *op. cit*., 201.

[203] Shōwa Statism was a political syncretism of Japanese extreme right-wing political ideologies, developed over a period of time from the Meiji Restoration. It is sometimes also referred to as Shōwa nationalism or Japanese fascism. This statist movement dominated Japanese politics during the first part of the Shōwa period (reign of Hirohito). It was a mixture of ideas such as Japanese ultra-nationalism, militarism and state capitalism, that were proposed by a number of contemporary political philosophers and thinkers in Japan.

reason for Japan's superiority."[204] Japan had decided to treat its conquests in the worst possible way, emulating what it considered the treatment dished out by European conquerors to Asians.

The Japanese ran a considerable number of internment camps throughout east Asia. Internees who ended up in these camps faced the most despicable conditions possible that included starvation, forced labor, and exposure to disease and extreme weather. In camps like Mukden, in the far north of China, prisoners of war slaved for companies such as Mitsubishi while fearing starvation and death-by-freezing in the harsh winters. In tropical camps such as the infamous Sandakan, in Borneo, internees faced tropical disease, jungle rot, relentless heat, infected water, and imprisonment in tiny enclosures exposed to the brutal midday sun. The Kempeitai, Japanese secret police, ran these camps, and would subject prisoners to beatings, torture, and death by beheading, bayoneting, and machine-gun execution.

Even the small South Pacific island of Nauru did not escape the horrors of the war. The massacre on Nauru was marked by extreme brutality. During their occupation of the island, the Japanese committed a string of atrocities, and a few stood out for their brutality. Following a successful raid by American bombers on March 1943, the Japanese beheaded and bayoneted five interned Australians in retaliation. That same year, the Japanese also forcibly deported more than 1,000 indigenous inhabitants as labor to other occupied islands to conserve rations. During their occupation, the Japanese exterminated the island's leper colony. Stowing the island's 39 lepers on a boat, the Japanese led them far out to sea and out of sight. Afterward, Japanese gun boats fired at the vessel, sinking it and killing all onboard.

On 18 March 1943 the Japanese destroyer *Akikaze* picked up German missionaries and Chinese civilians living in the South Pacific islands of Kairuru and Manu. En route to Rabaul, the ship received instructions to execute the entire group. After three hours, the Japanese had successfully killed all 60 of their passengers, including two children

---

[204] Wetzler, Peter. *Hirohito and War,* University of Hawaii Press, 1998, p.104; Bix, Herbert. *Hirohito and the Making of Modern Japan,* Harper Perennial, 2001, p.280.

whom they threw overboard while still alive.

The Sook Chin Massacre took place over several months in early 1942. After the British surrendered Singapore, the Japanese wanted to remove all possible resistance, especially among the Chinese living in the region. The notorious Japanese secret police Kempeitai initiated Operation Sook Ching (meaning "purge through cleansing") in February 1942. After interning and interrogating the city's entire Chinese population, the Kempeitai herded those they deemed as dangerous into military vehicles. They then transported them to the city's outskirts and executed them all. But Singapore was merely the first to be purged. This purging operation soon found its way into other parts of Malaya as well. The Kempeitai were especially merciless toward those in rural areas. They eliminated entire villages on mere suspicion of subversive activity. Although there were no official casualty estimates it is probable the between the police executed at least 5,000 according to Japanese sources. Allied estimates run as high as 100,000 executions.

The scale and savagery of the Rape of Nanking defies all explanation. In 1937, the Japanese invaded Nanking, the capital city of Nationalist China. The atrocities began in December of 1937 and continued into 1938. As many as 300,000 Chinese civilians were killed, and as many as 80,000 Chinese women were raped. The Japanese bayoneted infants, forced family members to rape one another, beheaded children, threw bodies into wells to poison the water supply, and buried civilians alive. It was the first of many similar massacres, though none took place on the same scale as that at Nanking.

The war crimes trials for Europe had preoccupied the Allies during much of the Second World War, but they had shown little interest in Japan. Finally, on 31 August 1945 Washington inquired of General Douglas MacArthur what his plans were for the punishment of the Japanese. MacArthur had already appointed Colonel Alva Carpenter chief of the War Crimes Branch and made him Judge Advocate General for the Pacific theater. There was no legal or political objection to MacArthur's establishment of rules and procedures for his area. Secretary of War Henry Stimson and his staff had argued as early as 1944 that the Allies should use military commissions to try war crimes as outlined in the War

Department's field manual. In March 1944 the Joint Chiefs of Staff had made it clear that they opposed any limitation of the powers of military commanders over war crimes prosecution. Thus, Colonel Carpenter's War Crimes Branch operated on the assumption that it could formulate constitutionally whatever rules and procedures it thought to be appropriate and would expect no outside objections. The Branch would not be restricted to using traditional American rules of evidence and procedure.[205]

For various reasons, few of them valid, many war crimes and Japanese war criminals went untried. For the most part the decisions to not prosecute were personally made by General MacArthur. After the war General MacArthur was determined to make Japan into a Western style democracy. The image of Japanese as barbarians who violated all the rules of warfare would work against this conversion. Many of the atrocities committed against American invading forces, and some atrocities committed against POWs were well known and thus could not be suppressed. But few Japanese were known to the public in the Allied democracies the way the German leaders were known. Few Americans or British could name any Japanese except the Mikado and Tōjō.

Among the true war criminals who escaped all trial and punishment were the following. Sasakawa Ryō'ichi was initially accused of war crimes and confined in Sugamo Prison, but was never tried, and then released. He joined the *yazuka*, Japan's mafia, working against the political left. Kodama Yoshio was also held in Sugamo Prison, but was soon released and was involved in various nefarious schemes. Colonel Naitō Ryō'ichi, the man who negotiated immunity for Unit 731 personnel, created a partnership with another Unit 731 figure General Kitano Masaji and both became multi-millionaires selling blood commercially.[206]

The Japanese record of war crimes in World War II was extensive. Most of the following figures are official, having been presented at the Tokyo War Crimes Trial. From the invasion of China in 1937 to the end of World War II, the Japanese military regime murdered near 3,000,000 to

---

[205]Lael, Richard L., *The Yamashita Precedent*. Wilmington, DE: Scholarly Resources, 1982, 59-64.

[206] Tanaka, Yuki, *Hidden Horrors: Japanese War Crimes in World II*. Boulder: Westview Press, 1996.

over 10,000,000 people, most probably almost 6,000,000 Chinese, Indonesians, Koreans, Filipinos, and Indochinese,[207] among others, including Western prisoners of war. This democide[208] was caused by a morally bankrupt political and military strategy. Military expediency and custom prevailed, based on a unique national culture. For example, Japanese held the view that those enemy soldiers who surrendered while still able to resist were criminals and this deserving of ill treatment. It also harbored the illusion that the Japanese were a race superior to all others and were led by a divine emperor, a descendant of the sun goddess.[209] Anything the Japanese chose to do was justified by this superior status.[210]

Of 41,862 known Chinese laborers sent to Japan, 2872 died in China before leaving, 600 on ships bound to Japan, 200 during their journey to the place of employment, and 6872 in Japanese workplaces. At least, 600,000 to 1,610,000 forced Asian laborers died. Of 300,000 Javanese, Tamils, Burmese, and Chinese who became forced laborers sent to build the Burma-Thailand railroad in 1942 to 1943 more than 60,000 died or were killed. Some 5,400,000 Koreans were conscripted for labor beginning in 1939 with at least 67,000, and perhaps as many as 80,000, dead between 1937 and 1945. Japan conscripted over 1,000,000 forced laborers from Manchuria with at least 100,000 dying in servitude.[211] As an example of Japanese brutality, some 250 Tamil men, women, and children, forced laborers in June 1943 in Nieke became infected with cholera and were confined to a single barracks. The Japanese set fire to the wooden building, immolating all within it.

Murder of citizens of occupied nations ranked high among Japanese war crimes. In the Philippines, where after the war American military teams made an effort to investigate Japanese massacres, about

---

[207]Kratoska, Paul H. *Asian Labor in the Wartime Japanese Empire.* Armonk, NY: Sharpe, 2005.

[208]Democide is the murder of any person or people by a government, including genocide, politicide, and mass murder.

[209] Dower, John, *War Without Mercy: Race and Power in the Pacific War.* New York: Pantheon, 1985.

[210] Lord Russell of Liverpool, *The Knights of Bushido: A Short History of Japanese War Crimes*. Berkeley Books, 1961.

[211] Landas, Marc, *The Fallen: A True Story of American POWs and Japanese Wartime Atrocities.* Wiley, 2004.

159,000, and possibly as many as 318,000, civilians were reported killed.[212] The number of civilians murdered in the 43 known massacres in China exceeds 1,000,000.[213] Of 130,000 Europeans interned in the Dutch East Indies, 30,000 died, among whom were 4500 women and 2300 children. At least 42,000, and possibly 80,000, Indonesians were massacred during the Japanese occupation. From the Indochinese -- Vietnamese, Laotians, and Cambodians -- population there is a high estimate of 575,000 dead. The Chinese living in Malaya suffered from the Japanese occupation with at least 37,000 of them being executed.[214] The overall Japanese democide in World War II, including Japanese democide in China included, provides a total of 3,056,000 to 10,595,000 with a likely mid-total of 5,964,000 people killed.[215]

The Allies created three classes of war criminals charged with ordering, supervising, and committing war crimes. The A class defendants were the principal decision and policy makers. From early 1946 to late 1948, some two hundred top Japanese civilian and military leaders were designated as Class A war criminal suspects; twenty-eight of them were tried at the Tokyo War Crimes Tribunal from early 1946 to late 1948. Of the fifty-five counts of crime listed in the indictment, the overwhelming majority fell under crimes against peace, fifteen counts under murder, and three under other conventional war crimes and crimes against humanity. Virtually all of these defendants bore some responsibility under the principle of command responsibility.

Classes B and C included those who ordered atrocities, allowed them to occur, or actually committed them. Between 1946 and 1951, the U.S. and other governments separately tried about 5,700 Japanese officers and enlisted men in some 2,240 separate trials throughout the Asia Pacific. Among them were over three hundred Koreans and Taiwanese employed by the Japanese. These so-called B and C class tribunals were concerned

---

[212] Escoda, Jose M., *Warsaw of Asia: The Rape of Manila*, revised ed. Quezon City, Philippines: Giraffe Books, 2001.

[213] Li, Peter, *Japanese War Crimes*. Transaction Publications, 2013.

[214] Kratoska, Paul, ed., *Southeast Asian Minorities in the Wartime Japanese Empire*. London: Routledge Curzon, 2002.

[215] Rummel, R.J., *Genocide and Mass Murder Since 1900*. Charlottesville, VA: School of Law, University of Virginia, 1997, chapter 3.

with those perpetrating or ordering conventional war crimes. Only a comparative handful of these crimes involved command responsibility as most crimes committed by this group concerned principally the individual.[216]

Other War Crimes

The Rape of Nanking,[217] was a mass murder and war rape that occurred during the six-week period following the Japanese capture of the city of Nanking, the former capital of the Republic of China, on December 13, 1937 during the Second Sino-Japanese War. Hundreds of thousands of Chinese civilians and disarmed soldiers were murdered by soldiers of the Imperial Japanese Army. Widespread rape[218] and looting also occurred. The commanding officer, General Iwame Matsui, had issued orders before the city fell instructing his troops to observe restraint in dealing with the civilian population. The man most clearly responsible for this atrocity was the uncle of Japanese Emperor Hirohito, Prince Asaka, a member of the imperial family. He escaped prosecution because General MacArthur earlier had granted immunity to all members of the royal family. According to one Japanese journalist embedded with Imperial forces at the time, "The reason that the [10th Army] is advancing to Nanking quite rapidly is due to the tacit consent among the officers and men that they could loot and rape as they wish."[219] Thus, it was Matsui whom the tribunal put on trial. Matsui could not possibly have prevented the events that occurred with the tacit approval of a member of the royal family and

---

216 Boister, N. & R. Cryer. *The Tokyo International Military Tribunal: A Reappraisal.* Oxford University Press, 2008.

217 Iris Change, Rape of Nanking. London: Penguin, 1998; Daqing Yang, "Convergence or Divergence? Recent Historical Writings on the Rape of Nanking" 104 *American Historical Rev.* (June 1999) 842-65.

218 For so-called comfort women, really Chinese women who were brutally raped, see Hicks, George, *Comfort Women: Japan's Brutal Regime of Enforced Prostitution in the Second World War.* New York: Norton, 1997; and David Andrew Schmidt, *Ianfu–The Comfort Women of the Japanese Imperial Army of the Pacific War: Broken Silence.* Lewiston, NY: Edwin Mellen Press, 2000; and Yuki Tanaka, *Japan's Comfort Women: Sexual Slavery and Prostitution During World War II and the U.S. Occupation.* London: Routledge, 2001.

219 *Japan Advertiser*, 7 December 1937. This newspaper was an American owned and edited English-language daily paper in Tokyo.

was thus powerless. Nonetheless, the tribunal found Matsui responsible for the criminal acts of the Japanese troops in Nanking. He was found guilty because he must have known of these criminal acts.[220] Historians have estimated that 250,000 to 300,000 people were killed. The International Military Tribunal of the Far East estimated more than 200,000 casualties in the incident.[221] The Nanking Massacre is usually said to have occurred over a six-week period after the fall of Nanking, the crimes committed by the Japanese army were not limited to that period. Many atrocities were reported to have been committed as the Japanese army advanced from Shanghai to Nanking.

In 1937, the Osaka Mainichi Shimbun and its sister newspaper the Tokyo Nichi Shimbun covered a "friendly contest" between two Japanese officers, Toshiaki Mukai and Tsuyoshi Noda, both from Island troops, the Japanese 16th Division, in which the two men were described as vying with one another to be the first to kill 100 people with a sword before the capture of Nanking. From Jurong to Tangshan, cities in Jiangshu Province, China, Toshiaki Mukai had killed 89 people while Tsuyoshi Noda had killed 78 people. The contest continued because neither of them had killed 100 people. By the time they had arrived at Zijin Mountain, Tsuyoshi Noda had killed 105 people while Toshiaki Mukai killed 106 people. Both officers supposedly surpassed their goal during the heat of battle, making it impossible to determine which officer had actually won the contest. Therefore, according to the journalists Asami Kazuo and Suzuki Jiro, writing in the they decided to begin another contest, with the aim being 150 kills. The Nichi headline of the story of December 13 read "'Incredible Record' [in the Contest to] Behead 100 People—Mukai 106 – 105 Noda—Both 2nd Lieutenants Go Into Extra Innings." After Japan surrendered, Toshiaki Mukai and Tsuyoshi Noda were arrested, tried by the Chinese in a war crimes tribunal, convicted, and executed by firing squad in Nanking charged with being civilized public enemies.[222]

---

220 *Official Transcript of the International Japanese War Crimes Trials in the International Military Tribunal for the Far East*, 49: 815-16. Herein after *Japanese War Crimes Trials*.

221 Yamamoto, Masahiro, *Nanking: Anatomy of an Atrocity.* Westport, CT: Praeger, 2000.

222 Wakabayashi, Bob Tadashi, "The Nanking 100-Man Killing Contest Debate: War

Chemical and biological warfare had long since been outlawed. Japan used both in China. The one trial of a Japanese officer for use of chemical warfare involved Kajiura Ginjirō, Commander of the 231st Regiment, 39th Division. He was tried and sentenced to life in prison by a military tribunal in China in 1947 for using poisonous gas in combat. From December 25 to 31, 1949, the Soviet government tried twelve Japanese officers involved in biological warfare operations. The transcript of the trial included both photocopied documents and photographs that supported this charge.[223]

General Shunroka Hata, commanding forces over a prolonged period in China, was tried for numerous criminal acts perpetrated by troops under his command. The tribunal assumed that Hata must have known of the atrocities because of their huge scale. It was his duty, the tribunal reasoned, to know of such matters and then take all corrective measures necessary to correct them. When he failed to restrain his troops who were committing atrocities against Chinese civilians he could be held responsible as a war criminal.[224]

The case of General H. Kimura, commander of the Burma area, is one of the more unusual post-war prosecutions. Kimura had issued orders to his troops to treat POWs fairly and well and to generally restrain themselves. There was no question that he had given these two specific orders in a clear and unequivocal manner. His subordinates evidently ignored and circumvented the orders resulting in maltreatment of the prisoners and numerous criminal activities regarding the Burmese natives. The tribunal found him guilty of violating his command responsibility for failing to follow up and ascertain if his orders were being followed.[225] In both the Hata and Kimura cases the defendants were found to be culpable because they had failed to gather facts, read and study reports, and otherwise become aware of

---

Guilt Amid Fabricated Illusions," *Journal of Japanese Studies* 26: 2, [Summer 2000)]: 307–4.

[223] *Materials on the Trial of Former Servicemen of the Japanese Army, Charged with Manufacturing and Employing Bacteriological Weapons.* Moscow: Foreign Languages Publishing House, 1950.

[224] *Japanese War Crimes Trials*, 49: 784.

[225] Burnett, *Command Responsibility*, 114.

what was transpiring in the their respective commands.

Hisakazu Tanaka (1889-1947) was a general in the Imperial Japanese Army, and head of the Japanese occupation force in Hong Kong in World War II. His name is occasionally transliterated Tanaka Hisaichi. He was charged with command responsibility for the illicit execution of an American POW committed by a subordinate while he was absent from his post. The fact that he absented himself without providing adequate safeguards for such possibilities was sufficient for the trial judges to convict. Upon conviction an appeals court overturned the conviction because there was no evidence that he had "wrongful knowledge" of the crime and had no effective way of preventing it.[226]

Admiral Soemu Toyoda had served as Chief of the Naval General Staff from May until September 1945, having previously served as Commander of the Japanese Combined Fleet, May 1944 until May 1945. Tried, but no convicted, of war crimes, Toyoda should have become aware of atrocities committed under his command because of their great number. The tribunal wrote "that a reasonable man could have come to no other inclusion that that the accused must have known of the offenses, or of the existence of an understood and acknowledged routine for their commission." Under the principle of command responsibility if the accused "knew, or should by the exercise of ordinary diligence, have learned of the commission by his subordinates" of extralegal and criminal activities he could be himself held criminally responsible.[227]

Hirota vs. MacArthur

Baron Koki Hirota was a major player in Japanese politics in the 1930s, and was the only civilian sentenced to death by the IMTFE. A one-time Prime Minister of Japan and Foreign Minister during the earliest stages of World War II in Asia, Hirota was convicted on three of nine counts—for his role as one of the leaders, organizers, instigators, or accomplices in the formulation or execution of a common plan or "conspiracy . . . [to wage] . . . wars of aggression, and wars in violation of

---

[226] United Nations War Crimes Commission: Law Reports of Trials of War Criminals, 5:66.

[227] *U.N. War Crimes Commission: Law Reports,* 4: 118-19, *U.S. v Toyoda.*

international law;" for "waging a war of aggression against . . . . China;" and for deliberately and recklessly disregarding his duty to take adequate steps to prevent atrocities. Essentially, Hirota was really convicted for his complicity in the Rape of Nanking. Hirota was either derelict in his duty by not insisting before the cabinet that immediate action be taken to stop the Nanking atrocities, or he failed in other actions available to him to get the same results

Notwithstanding his complicity for the events at Nanking, Hirota's indictment was best described it as a borderline case. One of twenty-eight Class A defendants tried before the IMTFE, Hirota was one of the eleven who sought post-conviction relief in the courts of the United States. Hirota's was one of hundreds of similar applications for "original" habeas relief filed in the U.S. Supreme Court by Axis prisoners convicted of war crimes by American or Allied military tribunals. The Court denied every other original application without argument or an opinion, usually by a 4–4 vote with Justice Jackson recused because of his personal involvement at the Nuremberg Trials of Nazi war criminals.

There was no proof that Hirota, as foreign minister, had any legal authority to repress any war crimes, most importantly those that the Japanese perpetrated in Nanking. Reportedly, he had opposed the army's criminal activity and, in turn, was assured by the war minister and the army that no such incidents would ever occur again and that those who were responsible would be punished. The tribunal held that Hirota had a grave obligation to make his own independent investigation whether these assurances had been honored. If he found that the assurances were unkept he had a legal obligation to convince the whole cabinet to require that changes be made. Failing that he must resign from the cabinet.[228]

Hirota vs. MacArthur[229] was a decision by the Supreme Court of the United States, which was concerned with the jurisdictional aspect of war crimes trials and their apparent immunity to judicial review by U.S. Supreme Court. The case held that the courts of the United States have no power or authority to review, to affirm, set aside or annul the judgments

---

[228]Minear, Richard R., *Victors' Justice: The Tokyo War Crimes Trials*. University of Michigan, 1971, 71-72.

[229]338 U.S. 197 (1948).

and sentences imposed by the International Military Tribunal for the Far East. The motions for leave to file petitions for writs of habeas corpus were thus denied. The appeal to the U.S. Supreme Court was made following the imposition of death sentences against Koki Hirota and six other Japanese leaders tried for war crimes. The military tribunal set up in Japan by General Douglas Arthur MacArthur as the agent of the Allied Powers was not a tribunal of the United States, and the courts of the United States have no power or authority to review, affirm, set aside, or annul the judgments and sentences imposed by it on these petitioners, all of whom are citizens of Japan.

Justice William Douglas agreed, but expressed his concern in a concurring opinion about a total lack of judicial control over the international tribunal. His concern was that the high court had held that "no court of the United States to which the potential appellate jurisdiction of this Court extends has jurisdiction over this cause." He wrote, "Today, Japanese war lords appeal to the Court for application of American standards of justice. Tomorrow or next year, an American citizen may stand condemned in Germany or Japan by a military court or commission. If no United States court can inquire into the lawfulness of his detention, the military have acquired, contrary to our traditions, a new and alarming hold on us." Further, "It has never been deemed essential that the prisoner in every case be within the territorial limits of the district where he seeks relief by way of habeas corpus." By removing itself wholly from such cases the Supreme Court "would have to conclude that the United States Generals who have custody of petitioners are bigger than our government to hold that the respondent officials of the War Department have no control or command over them."[230]

The Supreme Court was satisfied that the tribunal sentencing these petitioners was not a tribunal of the United States. The United States and other allied countries conquered, and then occupied and controlled, Japan. The military tribunal sentencing these petitioners was set up by General MacArthur as the agent of the Allied Powers. It was well within the authority of the Allied High Command to create and constitute the war crimes tribunal.

---

[230] 338 U. S. 197 at 202ff (1948).

Mistreatment of POWs

One of the major Japanese war crimes was the mistreatment of prisoners of war [POWs] along with the impressment of civilian laborers in occupied lands. The overall number of POWs and internees killed was about 138,000. The prisoner of war death rate averaged nearly 29 percent. As many as 60,000 POWs were impressed to assist in building the Burma-Siam Railroad more than 25% died or were murdered. By comparison about 4% of British and American POWs died in German camps.

While few Americans knew the names of any of the Japanese hierarchy, there was one notable exception: Hideki Tōjō. His war crimes against prisoners of war (POWs) were unprecedented. It was Tōjō who had said that if a POW did not work, he would not eat. He knew of the Bataan Death March and the impressment of POWs in the building of the Burma-Siam Railroad. He was acutely aware that ill men were being driven out to work under conditions that would kill them. The death rate for POWs under Tōjō was appalling. Overall, 27% of all prisoners died, with the American rate at 34% highest of all nationalities. Some 33% of Australian and 32% of British POWs died under Japanese control, but less than 20% of the Dutch POWs died. By contrast, in Germany on the Western front the POW death rate was less than 4%.[231] Tōjō was tried by the International Military Tribunal for the Far East for war crimes and found guilty.

Those who wish a full and highly readable account of the fate of Allied POWs at the hands of Japanese captors are remanded to Gavan Daws, *Prisoners of the Japanese.*[232] This author provides both the statistical information and first-hand accounts of many POWs along with

[231] Daws, Gavan, *Prisoners of the Japanese*. Morrow, 1994, 360. See also *Japanese Prisoners of War*, eds. Philip Towle et al. London: Hambledon and London, 2000.

[232] Id. See also Moore B, Fedorowich K, eds. *Prisoners of War and Their Captors in World War II*. Oxford: Berg; 1996.

their assessments of various Japanese guards. He also relates how so many POWs risked, and often received, severe discipline for assisting their fellow prisoners. His descriptions of the starvation diets, terrible diseases, primitive medicine, and infestation by all sorts of creatures are compelling reading. Anyone who contests the meager punishment the Allies meted out to the barbarian guards and overseers must read this book.

The listing of Japanese atrocities perpetrated against Allied POWs herein is not meant to constitute a definitive enumeration of these horrid events. Indeed, it could be called at best a random sample of those crimes. But for our purposes here it should suffice to show that, for reasons of their maltreatment of Allied POWs alone, the Japanese violated the ultimate moral order and by any standard of justice should have lost the Pacific War.

Some of the Worst Crimes Against POWs

A site on the net called List Verse has a list of the ten most notable atrocities the Japanese perpetrated during World War II. Below are those that concerned prisoners of war.[233]

The Laha Airfield Massacre occurred on several days in February 1942. The Japanese killed more than 300 Australian and Dutch POWs, following the Japanese capture of the Indonesian island of Ambon. Reportedly it was an act of reprisal after the Allies destroyed one of their minesweepers. The Japanese selected prisoners at random and executed them by beheading and bayoneting near the island's airfield. They then repeated the process three more times during the month of February 1942. Australian military tribunal prosecuted more than 90 Japanese officers and soldiers after the war in one of the largest war crime trials in history. The tribunal sentenced four of the accused to death and handed out a range of sentences for the others. Unfortunately, the Allies never got to try Rear Admiral Hatakeyama who ordered this crime because he died of natural causes while awaiting his trial.[234]

The final successful Japanese raid by warships took place on 18 March 1944. The heavy cruiser *Tone* sank the British merchant vessel

---

[233]http://listverse.com/2014/05/06/10-japanese-atrocities-from-world-war-ii/

[234]https://dutcheastindies.webs.com/laha_massacre.html

Behar and took 108 prisoners in the process. Captain Haruo Mayuzumi relayed the good news of his ship's success to his superior Rear Admiral Naomasa Sakonju, expecting praise. Instead, the admiral berated the captain for bringing along useless prisoners. He ordered their execution. Mayuzumi appealed to his superior several times to spare the survivors. The admiral did not relent, and Mayuzumi carried out his orders. He divided the survivors into two groups composed of 36 and 72 members. The first contained the Behar's captain and other ranking personnel, who Mayuzumi transferred to a second ship, thus setting them free. However, he executed the second group. The Japanese beheaded them all and threw their bodies to the sea. The Allies hanged Sakonju while Mayuzumi received a seven-year imprisonment for his role in the incident.[235]

The most notorious submarine in the IJN was almost certainly the I-8. That ship is best remembered for sinking two Allied ships and for its terrible conduct afterward. On 26 March 1944, the submarine sank the Dutch freighter *Tsijalak* off the coast of Ceylon [now Sri Lanka]. The Japanese took 103 survivors onboard and massacred them with swords and sledgehammers. They then bound those still alive and left them on deck as the submarine dove below. Only five survived that ordeal. A few months later, the Japanese destroyed the U. S. cargo ship *Jean Nicolet,* repeating the same brutal treatment. The Japanese tortured and killed their prisoners by making them pass through a gauntlet of swords and bayonets before throwing their bodies overboard. The Japanese submarine later dived after spotting an Allied aircraft leaving 30 POWs still above deck. Only two dozen of the 100-plus prisoners survived.[236]

In February 1942, with the fall of Singapore to the Japanese imminent, sixty-five Australian Army nurses were evacuated on the steamer *Vyner Brooke.* In addition to the Australian nurses, the ship was crammed with over two hundred civilian evacuees and a few English military personnel. As the *Vyner Brooke* was passing between Sumatra and Borneo, Japanese aircraft bombed and strafed the overloaded ship and it sank quickly. Japanese aircraft strafed the lifeboats but some reached

---

[235]http://remorika.com/heavy-cruiser-tone/; also http://www.combinedfleet.com/Junyokan.htm

[236]http://military.wikia.com/wiki/Japanese_submarine_I-180

Bangka Island off the coast of Sumatra. Twelve Australian nurses were either killed in the attack on the ship or drowned in the sea, but fifty-three nurses reached Bangka Island. Wearing their Red Cross armbands, they were protected as non-combatants under convention of civilized nations. The nurses expected to be treated in a civilized manner by the Japanese when they reached shore. The Japanese herded the survivors at gunpoint into a small dirty building. Twenty-two Australian nurses were ordered by the Japanese to form a line and walk into the sea. Only one nurse survived the massacre. She survived imprisonment and gave evidence of the massacre at a war crimes trial in Tokyo in 1947.[237]

The Sandakan Death Marches involved mostly Australian POWs. Between 1942 and 1943, over 2,300 Australian and British POWs were sent to build an airstrip and service roads in Sandakan, a strategic base for the Japanese. The first 1500 prisoners who were mostly Australians from B Force arrived at the Sandakan POW camp on July 17, 1942 and endured the horrors of internment at the hands of the Japanese. Those caught smuggling medicine, food rations or making makeshift weapons and radios were beaten mercilessly and tortured in a variety of ways. The Kempei Tai would drive metal pins underneath the nails of the prisoner, singe their flesh with scolding flames and fill their mouths with water until their bodies began to bloat. Other times, the secret military police would stomp on the prisoners' stomachs and not stop until the swelling appeared. Determined to erase any evidence of the atrocities committed against their Allied captives, Japanese military officials ordered the POWS in early 1945 to march 260 kilometers through inhospitable virgin jungle from their camp to the Japanese garrison at Ranau, a rugged, hilly area that nestled beneath the shadow of Mt Kinabalu. Three 'Death Marches' that took place that same year. By January 1945, only 1900 POWs were left alive at Sandakan. Only the strongest including 470 Australian soldiers were selected to serve as porters on the arduous trek to northern Borneo. Any stragglers were killed by the Japanese so bodies were strewn along the trail. Of the 313 men who arrived in Ranau, all were too exhausted to march beyond the town. The second death march took place four months

[237]http://www.pacificwar.org.au/JapWarCrimes/TenWarCrimes/Banka_Massacre.htm
l

later and of the 536 prisoners who began the trek, only 183 reached Ranau's POW camp in late June. The 250 prisoners who were left behind at Sandakan were too ill to be moved. Initially, the Japanese intended to abandon the men knowing all would die of sickness and starvation. But as the Allied invasion known as 'The Borneo Campaign' quickly gained momentum the Japanese government did not want to leave any evidence of atrocities in their wake. The culling process began once again.

Selecting another group of 75 POWs, the third death march to Ranau commenced on June 9, 1945 though none of the men made more than 30 kilometers. Too feeble and weak to continue, they died where they fell. Some of the guards bludgeoned the men with a rifle butt; others stabbed them with their bayonets.[238]

Case of Masharu Homma

Masharu Homma (1887-1946) was born on Sado Island, in the Sea of Japan. He graduated from the Imperial Japanese Army Academy in 1907, and from the Army Staff College in 1915. Homma spent eight years as a military attaché in the United Kingdom, including combat service in France in 1918 with the British Expeditionary Force. From 1930-1932, Homma was again served as a military attaché to the United Kingdom, where his proficiency in the English language was useful. He was also assigned to be part of the Japanese delegation to the Geneva Disarmament Conference in 1932 and served with the Press Section of the Army Ministry from 1932-1933. He was given a field command again, as commander of the IJA 1st Infantry Regiment from 1933–1935, and was promoted to command the IJA 32nd Infantry Brigade from 1935-1936. In 1937, Homma served as aide-de-camp to Prince Chichibu, a brother of Emperor Shōwa. The two men made a diplomatic tour in Europe which ended in Germany. They met Adolf Hitler with whom the prince was impressed.[239] Homma then served as the commander of the Taiwan Army of the Imperial Armed Forces. Homma was appointed commander of the 27th Division, Japanese Imperial Army, in China from 1938 until 1940.[240]

---

[238]https://www.sandakanspirit.com/sandakan-death-march/

[239] Yoshimori Futara *et al. The Crown Prince's European Tour*. 1925.

[240] Bix, *op. cit.*, 162.

Homma was commander of the 14th Imperial Japanese Army which invaded the Philippines. He ordered his troops to treat the Filipinos as friends, and to respect their customs and religion. This liberal approach earned him the enmity of his superior, General Count Hisaichi Terauchi, commander of the Southern Army, who sent negative reports about Homma to Tokyo. There was also disloyalty and subversion within Homma's command led by a small group of subordinates, notably Colonel Tsuji Masanobu. In Homma's name, they sent out secret orders which ran contrary to his explicit orders and policies.

In conquering the Philippines rather than waste his men in furious frontal assaults, he tried to outmaneuver the American forces. His superiors then claimed that he had been "contaminated" by Western ideas regarding the conservation of the lives of his men. Deterioration in the relationship between Homma and the general staff led to the removal of Homma from command shortly after the fall of Corregidor, and he was thereafter only nominally commander of the 14th Army.

After the surrender of Japan, the American authorities arrested Homma. General MacArthur ordered his extradition to the Philippines in order that he could be tried by an American court martial. Homma was convicted of the actions of his men during the march rather than having a direct hand in the actions themselves, that is, command responsibility. It is far from clear whether Homma ordered, or even was aware of, the atrocities that occurred during the Bataan Death March. However, it is clear that he lacked the administrative expertise and ability to adequately delegate authority and control his men which led to the atrocities. After American-Filipino forces surrendered the Bataan Peninsula, Homma turned logistics of handling the estimated 25,000 prisoners to Major-General Yoshitake Kawane. Homma had publicly stated that the POWs would be treated fairly. Kawane formulated a plan to transport and march the prisoners to Camp O'Donnell, which Homma approved. However, the plan was severely flawed.[241]

Homma was indicted on 48 counts of violating the international

---

[241] Norman, Michael, and Norman, Elizabeth M. *Tears in the Darkness: The Story of the Bataan Death March and Its Aftermath.* New York: Farrar, Straus and Giroux, 2009.

rules of war, but during this first meeting with his lawyers the general said he was pleading *not guilty* to all of them. As the commander of the 14th Imperial Army he was "morally responsible," but he said he neither knew about nor condoned—let alone ordered—any of the crimes for which he was now being charged. Of all the charges, he seemed to understand that those associated with the Bataan Death March would be the hardest to defend against. And yet Homma appeared to have only a vague notion of what this incident was supposed to have been. He said the very first time he'd heard the term was shortly before being taken into American custody, when several reporters asked him about his role in the atrocity.

In his defense at his trial, Homma also claimed that he was so preoccupied with the plans for the Corregidor assault that he had forgotten about the treatment of the prisoners, believing that his Kawane was properly handling the matter. He claimed that he did not learn of the atrocity until after the war. But there were other charges. Atrocities continued at Camp O'Donnell and at Cabanatuan after the arrival of the POWs. Homma also bore responsibility for bombing Manila after it had been declared an open city. Troops under his command refused to grant quarter to troops on Corregidor.[242]

Homma was convicted by the U.S. military tribunal for war crimes in the Philippines, including the Bataan Death March, and the atrocities at O'Donnell and Cabanatuan which followed. As the commander of the 14th Area Army, it was his duty to ensure that the prisoners were properly cared for. He was responsible for ensuring that the logistics were ready to take care of the prisoners. He failed in his command responsibility to ensure that his troops behaved in a legal and humane fashion. All of these were tasks incumbent upon him based upon his position. Having failed, he was thus legally responsible and accordingly could be held accountable. As a matter of international law, this was actually a new interpretation of command responsibility. The first person who had been convicted on its basis had been Lt. General Tomoyuki Yamashita.

Homma's chief defense counsel, John H. Skeen Jr., stated that it was a "highly irregular trial, conducted in an atmosphere that left no doubt

---

[242] Sides, Hampton "The Trial of General Homma," American Heritage, February/ March 2007.

as to what the ultimate outcome would be." Associate Justice Frank Murphy of the U.S. Supreme Court protested the verdict, stating: "Either we conduct such a trial as this in the noble spirit and atmosphere of our Constitution or we abandon all pretense to justice, let the ages slip away and descend to the level of revengeful blood purges."

Homma's wife appealed to General MacArthur to spare his life but he denied her pleas. However, MacArthur, as a small token to the woman, MacArthur ordered Homma to be shot, rather than hanged as Yamashita was. He was not allowed to wear the uniform of his rank as general. Homma was executed by firing squad by the Filipino and American forces on 3 April 1946 outside Manila.[243]

Yamashita Case

Tomoyuki Yamashita (1885-1946) was an Imperial Japanese Army general during World War II. He was most famous for conquering the British colonies of Malaya and Singapore. On 6 November 1941, Yamashita was put in command of the Twenty-Fifth Army. On 8 December, he launched an invasion of Malaya, from bases in French Indochina. In the campaign, which concluded with the fall of Singapore on 15 February 1942, Yamashita's 30,000 front-line soldiers captured 130,000 British, Indian, and Australian troops, the largest surrender of British-led personnel in history. He became known as the "Tiger of Malaya." The first of the atrocities with which the Allies charged Yamashita followed. The campaign and the subsequent Japanese occupation of Singapore included war crimes committed against captive Allied personnel and civilians. The two most notable war crimes were the Alexandra Hospital and Sook Ching massacres.[244]

At about 13:00 hours on 14 February, Japanese soldiers advanced towards the Alexandra Barracks Hospital. A British lieutenant, acting under a white flag, approached the Japanese forces but was bayoneted and killed. After the Japanese troops entered the hospital, a number of patients,

---

[243] Piccagallo, Philip R. *The Japanese on Trial: Allied War Crimes Operations in the East, 1945-1951*. University of Texas Press, 1980.

[244] Havers, R. P. W., *Reassessing the Japanese Prisoner of War Experience: The Changi POW Camp, Singapore, 1942–45*. London: Routledge Curzon, 2003.

including those undergoing surgery at the time, were killed along with physicians, surgeons, nurses, and orderlies. Approximately 200 male staff members and patients remained alive, but were rounded up and bound. The following day this group was assembled, many of which were walking wounded. These men were ordered to walk to an industrial area. Anyone who fell was bayoneted. The Japanese guards forced the men into a series of small, badly ventilated rooms and were imprisoned overnight without water. Some died during the night as a result of their treatment. Those who remained were bayoneted the following morning. The Japanese did not consider those who were weak, wounded, or who had surrendered to be worthy of life. Many who had not been imprisoned in the tiny rooms in the industrial area were systematically taken away in small groups and killed with machetes and bayonets. This carnage continued for 24 hours, leaving 320 men and one woman dead. There were only five known survivors.[245]

Pursuant to Japan's unconditional surrender to the United States at the end of the war, General Yamashita also surrendered to United States troops present in the Philippines and immediately became a prisoner of war. He was detained by the United States Army in the Commonwealth of the Philippines. Upon his arrest as a prisoner of war, General Yamashita was charged by the Army's Judge Advocate General's Department with violations of the law of war. Forces under his command engaged in a "deliberate plan to massacre and exterminate a large part of the civilian population of Batangas Province as a result of which more than 25,000 men, women, and children all unarmed noncombatant civilians, were brutally mistreated and killed." Troops under Yamashita's command also had massacred POWs at Palawan. Because General MacArthur charged Yamashita only with war crimes, the MacArthur's Judge Advocate General could proceed without consulting the War Department. The prosecution could proceed without being hampered by traditional requirements for the admissibility of evidence and rules of procedure.[246]

---

245 Scott, James M. *Rampage: MacArthur, Yamashita, and the Battle of Manila.* New York: Norton, 2018. The book contains a balanced account of Yamashita's trial and execution.

246 Feldhaus, J. Gordon, "The Trial of Yamashita," *South Dakota Bar Journal*, 15: 181-93 [1946]; Ganoe, John, "The Yamashita Case and the Constitution" *Oregon Law Review*, April 1946, 143-58.

Direct testimony established his approval of execution of hundreds of suspected guerrillas without trial, and of thousands more without any kind of fair trial procedure. It also established that his headquarters were within walking distance of POW camps where numerous atrocities occurred. Certainly, he had actual knowledge of criminal misconduct by his subordinates and that was sufficient under World War II standards for failure to maintain command.

Upon his conviction, General Yamashita filed a writ of *habeas corpus* with the Supreme Court of the Commonwealth of the Philippines. In his writ, General Yamashita challenged the jurisdiction of the military commission, asserted that he did not commit a violation of the law of war, and claimed that he was denied a fair trial under the United States Articles of War, the Geneva Convention, and the United States Constitution. The Supreme Court of the Commonwealth of the Philippines denied his writ in total.

The General then appealed to the United States Supreme Court. In the Supreme Court's decision, rendered in February 1946, the Court articulated a clear standard for military commanders with respect to the actions of their subordinates. In responding to General Yamashita's assertion that he did not personally participate in or order the commission of these offenses, the Court described the heart of the charge as being "an unlawful breach of duty by [General Yamashita] as an army commander to control the operations of members of his command by 'permitting them to commit' the extensive and widespread atrocities." The Court recognized that international law, through the law of war, "presupposes that [violations of the law of war] are to be avoided through the control of the operations of war by commanders who to some extent are responsible for their subordinates." The Court believed that absent such a duty upon commanders, nothing would prevent occupying forces from committing atrocities upon the civilian population. The Court held that Yamashita was, by virtue of his position as commander of the Japanese forces in the Philippines, under an "affirmative duty to take such measures as were within his power and appropriate in the circumstances to protect prisoners of war and the civilian population."[247] Chief Justice Stone wrote,

---

[247] *In Re Yamashita,* 327 U.S. 1 (1946).

> The question then is whether the law of war imposes on an army commander a duty to take such appropriate measures as are within his power to control the troops under his command from the prevention of specified acts which are violations of the law of war and which are likely to attend the occupation of hostile territory by an uncontrolled soldiery, and whether he may be charged with personal responsibility for his failure to take such measures when violations result. . . . It is evident that the conduct of military operations by troops whose excesses are unrestrained by the orders or efforts of their commander would almost certainly result in violations which it is the purpose of the law of war to prevent. Its purpose to protect civilian populations and prisoners of war from brutality would largely be defeated if the commander of an invading army could with impunity neglect to take reasonable measures for their protection. Hence the law of war presupposes that its violation is to be avoided through the control of the operations of war by commanders who are to some extent responsible for their subordinates.[248]

Other Crimes Against POWs

The Palawan Massacre occurred in anticipation of an American invasion of that island. Among the American prisoners remaining in the Philippines were 346 men were sent 350 miles on 1 August 1942, to Puerto Princesa on the island of Palawan on the western perimeter of the Sulu Sea, to build an airfield for their captors. The brutal treatment they received at the hands of their Japanese guards was always the same. The men were beaten with pick handles, and kickings and slappings were regular daily occurrences. Prisoners who attempted to escape were summarily executed. Daily each of the prisoners received a mess kit of wormy Cambodian rice and a canteen cup of soup made from camote vines boiled in water. Prisoners who could not work had their rations cut by 30 percent. When six American POWs were caught stealing food in December 1942, they were tied to coconut trees, beaten, whipped with a wire and beaten again with a wooden club 3 inches in diameter. After this they were forced to stand at attention while a guard beat them unconscious, after which the prisoners were revived to undergo further beatings. A Japanese private named Nishitani punished two Americans, who were

---

[248]Id. at 14-16.

caught taking green papayas from a tree in the compound, by breaking their left arms with an iron bar.

On December 14, Japanese aircraft reported the presence of an American convoy which the Japanese thought was destined for Palawan. All prisoner work details were recalled. When the Japanese spotted two American Lockheed P-38 Lightning fighter aircraft the POWs were ordered into the air raid shelters. After a short time, the prisoners re-emerged from their shelters, but Japanese Lieutenant Yoshikazu Sato, whom the prisoners called the Buzzard, ordered them to remain in the area. A second alarm at 2 p.m. sent the prisoners back into the shelters, but 50 to 60 Japanese soldiers Sato' doused the wooden shelters with buckets of gasoline and set them afire with flaming torches. This the Japanese followed by throwing hand grenades. The screams of the trapped and doomed prisoners mingled with the cheers of the Japanese soldiers and the laughter of their officer, Sato. As men engulfed in flames broke out of their fiery deathtraps, the Japanese guards machine gunned, bayoneted and clubbed them to death. Most of the Americans never made it out of the trenches and the compound before they were barbarously murdered.[249]

General Shigeru Sawada was commander of the Japanese 13th Army in China. Troops under Sawada's command executed 8 Americans, refusing to accord them status as prisoners of war. Specifically, Okada Ryuhei and Lieutenant Wako Yusei, were both under Sawada's orders at Kiangwan Military Prison, Shanghai, China. Eight United States airmen, after taking part in a bombing raid on a Japanese steel mill, an oil refinery and an aircraft factory on 18th April, 1942, were forced to earth and captured by the Japanese and eventually held in Tokyo for about fifty-two days, during which time they were imprisoned in solitary confinement. There was evidence that, both during their period in Tokyo and previously, they were subjected to various forms of torture during interrogations. On 28 August 1942, after spending approximately 70 additional days at the Bridge House Jail, Shanghai, in small verminous and insanitary cells, all eight fliers were removed to the Kiangwan Military Prison, on the outskirts of Shanghai. At the time of their transfer, all the fliers were weak and underweight and one was very ill. On arriving, they were assembled in a

---

[249]https://www.historynet.com/american-prisoners-of-war-massacre-at-palawan.htm

room before several Japanese officers, who, they later learned, constituted their court-martial. The accused Wako and Okado were among the members of the court. Sawada's order was that they try as criminals the eight American airmen. Four fliers were sentenced to death by a Japanese military tribunal on false and fraudulent charges that Sawada had the power to commute. Instead Sawada refused to remit and revoke such sentences, thereby causing the unlawful death of four of the fliers and the imprisonment of the others.[250]

One of the worst war criminals of all times was Lt.-Col. Masanobu Tsuji (1901-1961). If one named a scene of carnage and atrocity it was likely under Masanobu's command. He was a leading proponent of *gekokujo*.[251] Tsuji was born in the Ishikawa Prefecture and received his education at a military academy and then graduated from the Imperial War College. Tsuji served as a staff officer in the Kwantung Army from 1937 to 1939. His aggressive posture contributed to the Soviet-Japanese Border Wars, particularly theChangkufeng Incident[252] and the Nomonhan Incident.[253] During the World War II he served principally in Malaya, Burma, the Philippines, and Guadalcanal. He served as a staff officer under General Tomoyuki Yamashita. Known as a brilliant tactician he planned the defense at Guadalcanal, he personally led the last two assaults by the Japanese forces which attempted to expel the Americans from the island. Tsuji returned to Tokyo after these failures to urge the evacuation of the troops from Guadalcanal. He impressed the Emperor with his candid

---

250 *U.S. v Shigeru Sawada.* U.S. Military Commission, case number 25, trial of Lt.-Gen. Shigeru Sawada and 3 others, Shanghai, 15 April 1946.

251 "leading from below" or "loyal insubordination."

252 Also known as The Battle of Lake Khasan (July 29, 1938 – August 11, 1938) and also known as the Changkufeng Incident in China and Japan, was an attempted military incursion of Japanese occupied Manchukuo into the territory claimed by the Soviet Union. This incursion was founded in the beliefs of the Japanese side that the Soviet Union misinterpreted the demarcation of the boundary based on the Treaty of Peking between Imperial Russia and Manchu Empire and subsequent supplementary agreements on demarcation, and furthermore, that the demarcation markers were tampered with.

253 In August 1939 the Soviet Union and Japan fought the largest tank battle the world had thus far seen. Under General Georgy Zhukov, the Soviets won a crushing victory at the what the Soviets called the Battle of Khalkhin-Gol.

assessment of the military situation.[254]

He was involved in war crimes throughout the Pacific war including the massacre of Chinese civilians in Singapore, the executions of numerous surrendered POWs during the Bataan Death March.[255] He was personally on Bataan, informing his troops that the emperor desired that all white POWs be killed. Numerous POWs later claimed that Tsuji had himself killed various prisoners on the Death March. There were other war crimes in China. It was Tsuji who was actually responsible for the murder of physicians, nurses, and patients at the Alexandria Hospital in Singapore. Tsuji personally killed many Chinese soldiers, POWs, and civilians, and order the massacre of hundreds more by troops he commanded. He boasted that he carried in his body bullets from five different enemy nations.[256] Masanobu Tsuji was regarded as the most notorious Japanese war criminal to escape trial after the war.[257]

A recent historian has written,

> Colonel Masanobu Tsuji [was] a fanatic repeatedly wounded in action and repeatedly transferred by generals exasperated by his insubordination. Tsuji once burned down a geisha house to highlight his disgust at the moral frailty of the officers inside it. His excesses were responsible for some of the worst Japanese blunders on Guadalcanal. He was directly responsible for brutalities to prisoners and civilians in every part of the Japanese empire in which he served. In northern Burma, he dined on the liver of a dead Allied pilot, castigating as cowards those who refused to share his meal: The more we eat, the brighter will burn the fire of our hatred for the enemy.[258]

Following the capitulation of Japan in September 1945, Tsuji went into hiding in Thailand anticipating being charged with war crimes. When

---

[254] Budge, Kent G., "Tsuji Masanobu" in The Pacific War Online Encyclopedia. http://pwencycl.kgbudge.com/T/s/Tsuji_Masanobu.htm.

[255] Falk, Stanley, *Bataan: The March of Death.* New York: W. W. Norton, 1962; Donald Knox, *Death March: The Survivors of Bataan.* New York: Harcourt Brace Jovanovich, 1981. Other estimates of American deaths range from 650 to over 2,000; estimates of Filipino deaths range from 9,000 to 16,500.

[256] Tsuji, Masanobu, *Japan's Greatest Victory, Britain's Worst Defeat*. Trans. by Margaret E. Lake. Da Capo Press, 1997.

[257] Daws, *op. cit.*, 374.

[258] Hastings, Max, *Retribution: The Battle for Japan, 1944-45*. Knopf, 2008, 53.

it was clear that this not to be, he returned to Japan and wrote of his years in hiding.[259] That recollection became a best seller in Japan. In fact, his memoirs made him so famous that he later was elected a member of the Diet. In April 1961, he traveled to Laos but was never heard from again. He was presumed to have been a casualty of the Laotian Civil War, and was declared dead on July 20, 1968.[260]

[259]*Senko Sanzenri* literally lurking.
[260]"Masanobu Tsuji" in Wikipedia.

Food Scarcity

Food, or, rather, the shortage of it, played a role in driving Japan to war. American's dominance was not only a result of the enormous industrial production in the United States but also its abundance of food. A usually ignored, but central preoccupation, of the Japanese government was securing an adequate food supply. Every sector of Japan's war economy[261] relied on the food supply, yet Japan entered the Pacific War despite its inability to supply its needs for food. Total war placed an immense strain on the food system. Food was the fundamental basis for Japan's wartime economy. Many decisions were based, at least in part, upon considerations of supplying sufficient food.[262] Even before Japan attacked the United States on 7 December 1941 its government had introduced food rationing. Already in 1940, rice—the chief item in the Japanese diet—had been subjected to rationing due to bad harvests in the Japanese colony of Korea and the demands of the Japanese military in China (since 1937) and Southeast Asia (since 1941). Fish, the other dietary staple, had all but ceased to be distributed in some areas in 1944. Food supplies were so meager that the average Japanese citizen was living at or near starvation level. By 1945 the average civilian caloric intake of the average Japanese citizen was a meager 78 percent of the minimum needed for health and physical performance.

The relentless extraction of food from China and Korea, used in order to feed the Japanese army. and, occasionally, the Japanese homeland caused chronic hunger and malnutrition among the Chinese and Korean population. The number of nationalities affected by the seizure of food supplies rapidly multiplied as the Japanese conquests

---

[261]Cohen, Jerome B. *Japan's Economy in War and Reconstruction*. Minneapolis: University of Minneapolis Press, 1949.

[262]Collingham, Lizzie, *The Taste of War: World War II and the Battle for Food.* Penguin, 2012.

spread throughout Southeast Asia.

Even before the beginning of World War II Japan had imported a substantial amount of its food stuff, both from its colonies in Korea and Taiwan and also from other nations, especially of Southeast Asia. In April 1941, even before the Japanese government declared war on the United States, rationing had to be introduced. However, the state of Japanese agriculture in of itself was not really too bad on a basic level. There were several considerations which arose during the war which dealt major blows to the state of Japan's agriculture and food supply chains. By the last year of the war these factors created a society that was on the brink of disintegration caused by food shortages. Indeed, there were riots caused by food shortages by the time of the Japanese surrender. We shall now consider these factors.[263]

First there was the Japanese militarist command of the total economy, with allocation of food being one of the most basic and vital of controls. The Japanese military establishment had essentially taken total control of the Japanese economy. It had, among other things, made military procurement of goods the absolute priority; set price controls on many items, including basic food supplies; and had instigated total mobilization of the entire populace for war. Military procurement gave priority for food distribution to the military.

In theory, Japanese soldiers on campaign were entitled to the so-called *Assaku Koryo* (compressed ration) consisting of dried rice, pickled plums, dried fish, salt and vinegar. The individual components were delivered in tins or cellophane bags. Most of the foods were wrapped in waterproof paper. During the war, these foods were prepared by Japanese schoolchildren who were drafted into the patriotic effort. Theoretically, following the Russo-Japanese War of 1905, the Japanese had adopted Western style field kitchens adapted to the needs and diet of their soldiers. However, information relayed by U. S. military intelligence showed that this was rarely the case in southeast Asia. Field kitchens were rarely encountered in the Pacific Theater war zones. There the Japanese island

---

[263] Van Royen, W., and Bowles, O. *Atlas of the World's Resources,* Volume II: *The Mineral Resources of the World.* New York: Prentice-Hall, Inc. for the University of Maryland, 1952.

garrisons typically did not have proper field kitchens cooking for them, and as such, received very irregular and limited food supplies along with inadequate equipment.

The home islands were heavily dependent on their slave colonies in Korea, Formosa, and China for basic foodstuffs. Before Pearl Harbor, imports from those areas accounted for thirty-one percent of Japan's rice consumption, ninety-two percent of its sugar, fifty-eight percent of its soy beans, and forty-five percent of its salt.

Military orders allocated all industrial products and resources, such as energy and fuel for use in mechanized agriculture and transport of food products, whether by ship or railroad. Military prioritization was a fact because Japan had originally imported many of these items. At this time Japan had not benefited from mechanized agriculture to the degree that was common in the United States and Western Europe. For the most part Japan's farms were essentially tenant based, with one landlord who employed a large number of farmers working the land. This produced a situation commonly called *latifundium* where a rural proletariat lived in desperate economic circumstances. Overpopulation of the land necessarily created downward stress in yield. Where industry failed to absorb surplus population, agriculture did so. Labor intensive agricultural production, notably in rice production, frustrated efforts to introduce labor-saving technology.

By the 1930s the Japanese military was drafting more and more men, in addition to redirecting other workers into military related activities, such as production of military hardware and vehicles. Thus, there were fewer and fewer hands to do the agricultural work that required substantial human effort. Increasingly, the Japanese politico-military machine pulled in more women, older men, children, and middle-class who would never have been involved in farming were it not for the food crisis. Some historians have estimated that as many as two-thirds of all IJA deaths suffered in the Pacific War were the result of illness and starvation.[264]

Fish, historically an important part of the Japanese diet, were also

---

[264] https: // www.quora.com / How-strong-was-a-single-Japanese-soldiers-combat-capability-in-World-War-Two

more difficult to get, because fishing vessels were less able to operate due to a lack of fuel. Additionally, there was a very good chance the fishing vessels would be attacked and sunk, just as with any other Japanese merchant shipping. Price controls on food have historically had the effect of reducing already short supplies. Shortages also meant that the black market became a main source of food for much of the Japanese population. The urban population was steadily losing weight and around a quarter of townspeople were suffering from malnutrition. Tuberculosis, beriberi, digestive, skin and vitamin-deficiency diseases were rife. The birth-rate had fallen and infant mortality had risen.[265]

Even before the beginning of World War II, Japanese merchant shipping was in short supply and so Japan relied heavily on foreign merchant fleets to assist in supplying its needs, including for food. Some estimates have projected a shortage of much as 50% of domestic shipping requirements. Of course, virtually all foreign cartage ended upon the commencement of hostilities. Imports of any kind from any foreign or colonial source became extremely difficult by 1945.[266]

Operation Starvation was a mission initiated by Admiral Chester Nimitz who wanted his naval operations augmented by an extensive mining of Japan itself conducted by the air force. While General Henry H. Arnold felt this was strictly a naval priority, he assigned General Curtis LeMay to carry it out.[267] LeMay[268] assigned one group of about 160 aircraft of the 313th Bombardment Wing to the task, with orders to plant 2,000 mines in April 1945. The mining runs were made by individual B-29 bombers at night at low altitudes.[2] Radar provided mine release information. The 313th Bombardment Wing received preliminary training

---

[265] Vance, L. M. "Useless Eaters" *Lew Rockwell Reports,* 17 December 2013. https:// www.lewrockwell.com / 2013 / 12/ laurence-m-vance/wwii-starvation-of-useless-eaters/

[266] Johnston, Bruce F., Mosaburo Hosoda, and Yoshio Kusumi. *Japanese Food Management in World War II.* Stanford University Press, 1953.

[267] Caldwell, Hamlin A., Jr., "Air Force Maritime Missions," *United States Naval Institute Proceedings,* October 1978.

[268] LeMay, Curtis E. *Mission with LeMay.* Doubleday, 1965. Partially ghost-written and edited by McKinley Kantor, this is LeMay's story of his command in World War II. See also LeMay's *Superfortress: The B-29 and American Air Power.* McGraw-Hill, 1988. Partially ghost-written and edited by Bill Yenne.

in aerial mining theory while their B-29 aircraft received bomb-bay modification to carry mines.[269]

Beginning on 27 March 1945, 1,000 parachute-retarded influence mines with magnetic and acoustic exploders were initially dropped, followed by many more, including models with water pressure displacement exploders. This mining proved the most efficient means of destroying Japanese shipping during World War II. In terms of damage per unit of cost, it surpassed strategic bombing and the United States submarine campaign.[270]

By the end of the war, most of the major ports and straits of Japan had been repeatedly mined, severely disrupting Japanese logistics and troop movements for the remainder of the war. So effective was this operation that the Japanese had to abandon 35 of 47 essential commercial shipping routes. Shipping through Kobe alone declined by 85%, going from 320,000 tons in March to only 44,000 tons in July. Operation Starvation sank more ship tonnage in the last six months of the war than the efforts of all other sources combined. The Twentieth Air Force flew 1,529 sorties and laid 12,135 mines in twenty-six fields on forty-six separate missions. Mines sank or damaged 670 ships totaling more than 1,250,000 tons.[271]

Following the end of hostilities, American analysts reported in July 1946 to the United States Strategic Bombing Survey that it would have been more efficient to combine the American anti-shipping submarine effort with land- and carrier-based air power to strike harder against merchant shipping and begin a more extensive aerial mining campaign earlier in the war. This almost certainly would have starved Japan, forcing an earlier end to the war

Just as farming was accomplished with the application of intensive

---

[269] Hansell, Haywood S., Jr. Strategic Air War Against Japan. U.S. Government Printing Office, 1980. Hansell was a major general in the U.S.A.F.

[270] Parillo, Mark P. *The Japanese Merchant Marine in World War II.* Annapolis: Naval Institute Press, 1993. This work chronicles the fate of the Japanese merchant fleet, complete with useful charts and graphs.

[271] Craven, Wesley F.; Cate, James L., eds. *The Pacific: Matterhorn to Nagasaki, June 1944 to August 1945, Vol. V of The Army Air Forces in World War II*, University of Chicago Press, pp. 662–73.

labor, so was much domestic industrial production was accomplished by labor-intensive cottage industry. Much of that Japanese small-scale industry was destroyed by allied bombing.[272] Indeed, allied strategic bombing was undertaken specifically to stop Japan's war capabilities, such as by destroying aircraft and armament factories. However, the Japanese were able to rebuild many such factories simply because their industrial enterprises were nowhere near as sophisticated as the ones in the United States. It took the destruction of small-scale cottage industry, accomplished by massed fire-bombings, to cripple the Japanese industrial productivity.

The overpowering emphasis on war-related production had an obvious deleterious effect on the production of agricultural tools, farm machinery, fertilizer, and even spare parts for these machines. The destruction of distribution centers, logistical channels, warehouses, and so forth, all precluded the application of such vital items when they were produced. These factors stressed the Japanese economy as a whole, and food distribution in particular. While Japanese rice fields were typically not bombed, only because it was not cost-effective to do so, the ability to transport that food into the cities became a major concern. It must be noted that by late 1944, the United States actually had begun an aerial campaign designed to destroy much agricultural production. Crop failures compounded by the complete destruction of the merchant marine pushed the country to the brink of open famine.

The Allied Submarine campaign was especially effective. Once viable torpedoes were allocated the American campaign against the Japanese merchant marine proved incredibly useful in depriving the Japanese home islands of all raw materials, fuel, and agricultural products. As we had seen, the Japanese merchant marine was already over-burdened at the beginning of the war. Allied submarine warfare combined with the mass mining of Japanese ports essentially shut down Japan's naval shipping even further. This not only stopped the import and export of colonies but also hampered internal transport to areas historically sustained by extensive commerce.

---

[272] Allen, Thomas B. and Polmar, Norman. *Code Name Downfall: The Secret Plan to Invade Japan*. Simon & Schuster, 1995.

In addition to the merchant ships destroyed, U. S. submarines sank 700,000 tons of naval ships, which was approximately about thirty percent of the total lost, including eight aircraft carriers, a battleship and eleven cruisers. America deployed a total of 288 submarines throughout the war, including those stationed in the Atlantic Ocean. Altogether 52 submarines were lost with 48 being sunk in the Pacific theater. American submariners, who comprised only 1.6% of the Navy personnel, suffered the highest loss rate in the U.S. Armed Forces, with twenty-two percent killed.

The end result was that the only method to move food around was basically on land routes, and the lack of motorized vehicles meant that more often than not food would be transported by horse and buggy, severely limiting the speed and quantity of food which could be transported. The Allied policy of economic isolation and strangulation had been so effective that it had sent most of the Japanese merchant marine to the bottom of the ocean by mid-1945, choking off supplies to the home front as well as the war front. In the Southeast Asian and Pacific theaters, starvation became a major cause of death among the Japanese infantry.

Starvation threatened many Japanese garrisons on Pacific islands. Allied submarines had prevented resupply of many isolated posts. If there was not enough local food the soldiers were supposed to grow it themselves. On large islands, such as Rabaul, where a large Japanese military force trapped, troops were able to sustain themselves on near-starvation diets, until the end of the war by farming. Many of the troops had been farmers before conscription. Before embarking for Malaya in 1941 soldiers were instructed to use ingenuity, and if food were short, they should supplement their diet with anything that came to hand, including wild grasses. Several Japanese commanders, as on Wake Island, claimed they had executed captured Americans because they did not have sufficient resources even to feed their own troops. Most Japanese troops, accordingly, were left to fend for themselves combining locally available resources with whatever supplies that made it through Allied blockades. Their situation varied greatly depending on where they found themselves. Rabaul, with a huge isolated Japanese garrison, happened to be in an area with fertile soil surrounded by rich fishing grounds. Relatively speaking, troops there did well, even if they had to eat mostly unfamiliar foodstuffs.

A few other places like Truk enjoyed relatively decent farming and fishing resources. But most islands were hardly fertile, and their garrisons went hungry, so starvation was constantly a threat. Many other Central Pacific islands that were bypassed, such as Jaluit, Milli, and Maloelap, however, endured terrible conditions, because these islands had hardly enough soil to sustain more than a few palm trees. Additionally, the high seas that surrounded them had few edible and catchable fish. There were, at times, several thousand mouths to feed where there had been at best a few hundred natives or, in some cases, nobody at all before. Cannibalism was apparently more common than the Japanese acknowledged.[273]

Guadalcanal became known as Starvation Island. Each new Japanese unit deployed met the starving remnants of the previous force. One Japanese commander estimated that 15,000 Japanese soldiers had starved to death on Guadalcanal against only 5,000 had been lost in combat. One soldier, Sadao Suzuki, reported having fled into the island's jungles for six months and eaten whatever he could find. Said Suzuki, age 97 in 2015, "There weren't even any frogs, snakes or mice to eat." He boasted that the troops were highly motivated when first they landed on the island on 27 August 1942. But they quickly became short of food as they had been sent to the island on the assumption that they could take food from captured Allied forces. With battlefronts stretched out from Asia to the Pacific, Suzuki reported that Japan could not maintain its supply lines, making starvation the main enemy for Japanese soldiers on Guadalcanal. "Lots of soldiers died without food and water," Suzuki said. According to the official Japanese estimate, Japan sent a total of 30,000 soldiers to Guadalcanal, of which only 10,000 returned. Three-quarters of the 20,000 dead were noncombat fatalities, with starvation and illness key causes of death.[274]

Another Japanese commander, this one in New Guinea, complained that his enlisted men concentrated more in locating food than in defeating the Australian army. In many campaigns, commanders noted

---

[273]https://www.quora.com/What-did-the-Japanese-soldiers-eat-during-World-War-Two

[274]https://www.japantimes.co.jp/news/2015/05/26/national/history/ex-soldier-recalls-guadalcanal-island-death/#.XKDqd3dFx9A. See also https:// mainichi.jp/english/articles/20171208/p2a/00m/0na/005000c

that malnutrition was the principal reason for defeat. Lack of food, they concluded, had led to a decrease in morale and the breakdown of military discipline. Soldiers had become preoccupied with locating their next meal so that it was difficult to motivate them. Fighting was most fierce when the soldiers tried to hold on to territory where they controlled native gardens. Some officers noted that during the summer of 1944, many soldiers were reduced to eating *sacsac*, a tasteless, brown starch made from sago palms. In the Sepak River area the men were pleased to find dried grasses to eat.

Japanese troops practiced cannibalism on enemy soldiers and civilians in the last war, sometimes cutting flesh from living captives, according to documents discovered by Toshiyiki Tanaka, a Japanese academic in Australia. Cannibalism was the result of both gross shortage of food, and a desire to consolidate the group feeling of the troops. The documents Tanaka found concerning cannibalism included captured Japanese army memos as well as sworn statements prepared by Australian soldiers for war crimes investigations. Tanaka amassed at least 100 documented cases of cannibalism of Australian and Indian soldiers as well as Asian forced laborers in New Guinea. He has also found some evidence of cannibalism in the Philippines. A Japanese general issued an order in December 1944 stating that "While troops were permitted to eat the flesh of the Allied dead, they must not eat their own." Several U.S. Army documents indicated that the Japanese on New Guinea ate each other, as well as members of the local population, Asians who had been brought to the islands as slave laborers, and even Allied soldiers. One statement by an Australian lieutenant describes how he found the remains of a number of bodies, including one 'consisting only of a head which had been scalped and a spinal column. In all cases, the condition of the remains was such that there can be no doubt that the bodies had been dismembered and portions of the flesh cooked,' concluded the statement. Another statement from an Australian corporal tells how he found the mutilated bodies of colleagues whom he had earlier helped to bury in Japanese- occupied territory.[275]

[275]Tanaka published his research in 1993 as *Hidden Horrors: Japanese War Crimes in World War II.* The book was first published in Japanese, with an English translation

One of the first to level charges of cannibalism against Japanese soldiers was Jemadar Abdul Latif of the Indian Army who was rescued by the Australians at Sepik Bay in 1945. He alleged that Indian POWs and local New Guineas were killed and eaten by Japanese. A Pakistani who was conscripted by the British just in time to be captured when Japan overran Singapore was taken to New Guinea. That man testified that in his area Japanese soldiers killed and ate one prisoner a day for about 100 days.[276] A British corporal said he saw flesh being cut from prisoners who were still alive. Indian Captain R U Pirzai and Subedar Dr Gurcharan Singh related that: "Of 300 men who went to Wewak with me, only 50 got out. Nineteen were eaten. A Jap doctor, one Lieutenant Tumisa, formed a party of three or four men and would send an Indian outside the camp for something. The Japs immediately would kill him and eat the flesh from his body. The liver, muscles from the buttocks, thighs, legs, and arms would be cut off and cooked."[277]

In the Philippines, after Allied landings, Japanese General Yamashita estimated that 400,000 of the 498,000 Japanese deaths were caused by starvation. Perhaps sixty per cent, or more than one million, of the total 1.74 million Japanese military deaths between 1941 and 1945 were caused by starvation and diseases associated with malnutrition.[278]

Although the Japanese made enormous progress early in their campaign in Burma, the high command of IJA had made little provision for supplying their troops, especially with food. Partly because monsoon rains made effective campaigning possible only for about half of the year, the Burma campaign was among the longest of the war. During the campaigning season of 1942, the Japanese had conquered Burma, driving British, Indian and Chinese forces from the country and forcing the British to retreat into India. By the time the Japanese had conquered Burma their

in 1996. The book focuses on Japanese atrocities Australian territory and deep in the jungles of Borneo and Papua, New Guinea where thousands of Australian, British, Pakistani and Indian POWs were massacred, and some cannibalized.

[276]Sharma, Manimugdha, "Japanese Ate Indian POWs, Used Them as Live Targets in WWII," *Times of India*, 11 August 2014.

[277] *Australian Courier-Mail*, 25 August 1945.

[278]https://www.lewrockwell.com/2013/12/laurence-m-vance/wwii-starvation-of-useless-eaters/

supply of food was nearly exhausted. One of the principal reasons, if not the single most important, reason, for the Japanese defeat in Burma was hunger.

In the campaign to retake Burma the Allies first secured the coastal areas. Then the Allies were free to construct airbases which could be supplied by sea once they captured the two offshore islands, namely Ramree Island and Cheduba Island. Cheduba, the smaller of the two islands, had no Japanese garrison but the Battle of Ramree Island continued for six weeks after the initial landings on 21 January 1944 by the 26th Indian Division. The survivors of the small but tenacious Japanese garrison withdrew from the island, suffering heavy losses which were largely due to starvation and diseases exacerbated by inadequate diet.

In the Battle of the Admin Box from 5 to 23 February, the Japanese concentrated on an area defended mainly by line of administrative and communication troops. Fortunately, these troops were supported by tanks with which the Japanese were unable to deal. Although the actual number of battle casualties was approximately equal, the result was a major Japanese defeat. They had failed to panic Allied troops and because the Japanese were unable to capture enemy supplies, they starved.

Also at the beginning of April 1944, the Japanese 31st Division under Lieutenant-General Kotoku Sato reached Kohima. The engagement lasted from 5 to 18 April. The Allied high command supplied its troops by air and eventually sent additional support. More Allied reinforcements arrived and renewed the offensive. But by now, the Japanese were at the end of their endurance. Their troops, particularly 15th and 31st Divisions, were starving, and with the addition of the monsoon, disease rapidly spread among the ill-fed troops.[279]

As its own food supplies ran out the Japanese Army in Burma counted on capturing Allied foodstuff. That was not to be. In one major engagement, the Battle of the Admin Box, waged from 5 to 23 February 1944, the Japanese were unable to capture enemy supplies as they had

[279] Allen, Louis. *Burma: The Longest War*. Dent, 2005; Bayly, Christopher and Harper, Tim. *Forgotten Armies: Britain's Asian Empire and the War with Japan*. Penguin, 2005; and McLynn, Frank. *The Burma Campaign: Disaster Into Triumph, 1942–45*. Yale Library of Military History. New Haven: Yale University Press, 2011.

planned, so they starved. The 33rd Division and Yamamoto Force made repeated efforts to defeat the Indian-British forces, but by the end of June they had suffered so many casualties both from battle and disease that they were unable to make any progress. The Japanese were at the end of their endurance. Their troops, especially the 15th and 31st Divisions, were starving. During the monsoon season, disease rapidly spread among them, exacerbated by starvation diets. The Imphal operation was finally broken off early in July, and the Japanese retreated. The Burma operation was the greatest defeat to that date in Japanese history. They had suffered 50,000 to 60,000 dead, with overall casualties exceeding 100,000. Most of these losses were the result of disease, malnutrition and exhaustion.[280]

A recent book demonstrates how famine in Japan, as well as in other nations on both sides of the conflict, was the result of mismanagement and neglect, particularly in India, Africa, and China. World-wide during World War II over 20 million people died from starvation or malnutrition and its associated diseases.[281]

These factors resulted in the creation of a highly unenviable situation for Japanese civilians. Domestic food supplies were becoming more and more difficult to acquire. In 1943 the food supply in Japan reached a critical turning-point. By mid-1943 even generally law-abiding citizens were resorting to the black market in order to buy food. By the end of 1943 the declining ration was beginning to cause serious malnutrition among the Japanese population. By August 1944, well before the systematic destruction of urban centers by air raids, thirty percent of the work force at the Mitsubishi factory in Tsurumi was diagnosed as suffering from beriberi, a disease caused by poor and insufficient diet. The Japanese civilian population was teetering on the edge of famine.

The urban population was steadily losing weight and around a quarter of townspeople were suffering from malnutrition. Although the Japanese prioritized food shipments, the average caloric intake fell twelve percent below the minimum daily requirement for the non-farming

---

[280]McLynn, Frank. *The Burma Campaign: Disaster into Triumph, 1942–45.* Yale University Press; 2011.

[281]Collingham, Lizzie, *The Taste of War: World War II and the Battle for Food.* Penguin, 2012.

population in 1944. Tuberculosis, beriberi, digestive, skin and vitamin-deficiency diseases were rife. The birth-rate had fallen and infant mortality had risen. By 1945, food shortages were disrupting the war effort and rending the social fabric. Factory absenteeism rose nation-wide, in large part because workers took time off to bargain and barter for food in the countryside. By July 1945, industrial absentee rates in major cities stood at forty percent or more, with the food problem cited as a major contributing factor.[282]

Losses of merchant vessels combined with the capacity loss due to convoying significantly reduced Japanese economic strength. Long before August 1945 it was clear to the Japanese leadership that the country was defeated, largely because of severe shortages of food, fuel, and raw materials. Indeed, Japanese industry had ground to a halt for lack of raw materials. Japan's war industries were unable to function without imports of steel, aluminum, iron ore and oil. The metal for Japanese fighter planes was so inferior that some engines cracked if they were flown at full throttle. This was rapidly becoming irrelevant as the country ran out of aviation fuel. Imports of 16 key materials fell from 20 million tons in 1941 to 10 million tons in 1944 and 2.7 million tons in the first 6 months of 1945. The specifics were impressive: Bauxite imports fell off by eighty-eight percent just between the summer and fall of 1944. In 1945, pig iron imports plunged by eighty-nine percent, pulp by ninety percent, raw cotton and wool by ninety-one percent, fats and oils by ninety-two percent, iron ore ninety-five percent, soda and cement fell by ninety-six percent, lumber by ninety-eight percent, and fodder decreased by ninety-nine percent. Not a single ounce of sugar or raw rubber reached Japan. Japanese oil imports fell from 1.75 million barrels per month in August 1943 to 360,000 barrels per month in July 1944. The Japanese Navy alone required 1.6 million barrels monthly to operate. After September 1943, only twenty-eight percent of the petroleum shipped from the southern regions reached Japan. In the last 15 months of the war the figure was nine percent. Rice production in 1945 was approximately 60

[282]https://www.reddit.com/r/AskHistorians/comments/4p34oh/how_serious_a_problem_was_starvation_in_japan_in/

percent of what it was in 1942.[283]

In the immediate aftermath of the war, starvation was a major problem as the collapse of the militaristic rule led to significant lawlessness in the immediate post-war period. Civilian food stockpiles in southern Honshu, the most vulnerable area agriculturally, would have been used up by late Spring 1946 because of the Allied blockade policy. This would almost certainly have prompted a mass-migration north in search of food elsewhere, basically dissolving the civil government in southern Honshu.[284]

After the war, Paul Nitze[285] of the United States Strategic Bombing Survey (USSBS) team argued that Japan was already yielding in August 1945, largely because of food shortages, and that surrender would only have been a matter of time.

The author of a recent [and long overdue] study of the impact of food scarcity on the Japanese war effort has come to the following conclusions:

> In April 1941, even before the Japanese government declared war on the United States, rationing had to be introduced. If there was not enough local food the soldiers were supposed to grow it themselves. In 1943 the food supply in Japan reached a critical turning-point. By mid-1943 even generally law-abiding citizens were resorting to the black market in order to buy food. By the end of 1943 the declining ration was beginning to cause serious malnutrition among the Japanese population. Before embarking for Malaya in 1941 soldiers were "instructed to use ingenuity, and if food were short they should supplement their diet with anything that came to hand, including wild grasses." The Japanese leadership was prepared to starve not only the aged citizenry, but also those who were too young to contribute to the war effort. The Japanese civilian population was "teetering on the edge of famine." Long before August 1945 it

---

[283]https://ghb67.wordpress.com/2014/01/12/operation-starvation/

[284]https:// forums.spacebattles.com / threads / starvation-in-japan-in-1945-without-peace.394913/

[285]Paul Henry Nitze (1907 – 2004) was an American statesman who served as United States Deputy Secretary of Defense, U.S. Secretary of the Navy, and Director of Policy Planning for the U.S. State Department. He is best known for being the principal author of NSC 68 and the co-founder of Team B. He helped shape Cold War defense policy over the course of numerous presidential administrations.

was clear to the Japanese leadership that the country was defeated if only because of scarcity of foodstuff at a level sufficient to sustain the population, without even considering the feeding of POWs. The urban population was steadily losing weight and about one quarter of townspeople were suffering from malnutrition. Tuberculosis, beriberi, digestive, skin and vitamin-deficiency diseases were rife. The birth-rate had fallen and infant mortality had risen. .[286]

[286] Collingham, L. *The Taste of War: World War II and the Battle for Food.* Penguin, 2012, pp. 271-313.

Medicine and the Japanese Military

American soldiers complained of a variety of diseases and conditions under the umbrella term, Jungle Rot. these infestations included a variety of terrible and mysterious fungal skin diseases that soldiers contracted in the dense rain forests that covered tropical islands. Malaria, dysentery, typhus, and a number of other ailments sidelined more soldiers of all nations that fought in the Pacific War than did the opposing armies. Japanese soldiers were subject to the same debilitating jungle diseases as Americans. The great difference lay in the simple fact that the Japanese soldiers did not have access to care of the quality of medical care and drugs that were readily available to Allied soldiers.

The Pacific Theater of World War II was fraught with many dangers for medical personnel. The Red Cross symbol, used by Allied medical personnel, was not one that the Japanese regularly treated with respect. Japan had not signed the Geneva Convention before the war and did not feel obligated to abide by the international rules of conduct toward medical personnel. American medical crews quickly learned to smear mud over the red and white symbol emblazoned on their tents, helmets, and trucks to prevent themselves from being more of a target than they already were. In some areas, such as New Guinea, American medics attached to the field hospitals were given target practice with M-1 carbines when some of their noncombatant colleagues were killed by the enemy. The Japanese medics generally used no such distinguishing markings. They gave no quarter to enemy medics and, in turn, expected none for Japanese medical personnel.[287]

---

[287]Condon-Rall, Mary Ellen and Albert E. Cowdrey. *United States Army in World War II: Medical; Service in the War against Japan.* Center of Military History, 1998.

Japanese Military Medicine

The history of the Imperial Japanese Army Medical Department began in 1870 when the Imperial Army sent thirteen students to Germany to study western medicine as practiced in the German Army. Always impressed by German military ways, Japan in 1877 established a medical school staffed with German professors, and opened at the Imperial University. Simultaneously, the Imperial Japanese Army sent a number of its officers to train in Germany. Japan then modeled its command structure along with its system of medical support in the style of the German Army.

At the beginning of the Second World War, the Imperial Japanese Army still maintained the same basic medical system that had been created during the Russo-Japanese War. Medical Department officers received command rank and status within in the army and had their own independent chain of command.

During the Second World War, some eight million men served at one time or another in the military. Japanese authorities themselves estimated one and a quarter million men died while in service. Of the wounded only 530,000 survived. A Japanese medical officer admitted to an American interrogator that one soldier was killed for every two wounded a ratio considerably worse than achieved by the Japanese Army less than forty years earlier.[288]

The Japanese code of honor, as extended to military service, held that a soldier must never surrender. Nothing was more dishonorable than being taken as a POW. Likewise, it was dishonorable to become ill. Sickness was shameful. Preserving his health was at least as important as maintaining one's weapon and effects. A true soldier did not choose to report to sick call. It was the national duty of a soldier to eat good food, drink clean water, and ingest such pills and supplements as he was issued. Boiling tea was nearly as important as adding chemicals to polluted water.

---

[288]Hawk, Alan J. "The Paradox of the Imperial Japanese Medical Department" Academia.edu/2898317. See also Allied Translator and Interpreter Section (ATIS), South West Pacific Area, "Enemy Publication No. 247, Operation of Reorganized Division Medical Units:" 4.2. See also Gabriel, Richard and Karen Metz, *A History of Military Medicine, Vol II, From the Renaissance through Modern Times,* New York: Greenwood Press, 1992, 225-226.3. ATIS. Research Report 83: Organization of Medical Units in the Japanese Army, 29 January 1944.

Each soldier was issued a small quantity of chlorine powder for water purification. A water purification kit consisted of three glass test tubes inside a small medicine tin. The kit could purify one liter of water. Creosote pills warded off dysentery, the only disease more common in the Japanese Army than beriberi.

The Kongo No Hong, the Bushido Code of March 1928, contained no provision for retreat or surrender. It promulgated only the offensive option. Even defense was minimized. A warrior was always of the attack. A soldier who was too ill to participate in an attack was of no use; he had failed his emperor by become disabled.

During the Second World War, information from Allied Headquarters in India indicated that the organization and general standard of Japanese medical services for the evacuation and treatment of casualties was comparable to that provided by the Western Allies. The Japanese paid strict attention to developing and maintaining precautions for the prevention of those diseases which Japanese soldiers were most likely to contract. The Japanese did not overlook the damage which the incidence of such diseases could inflict on the morale and efficiency of an army in the field.

In addition to units of the army medical corps there were detachments of medical officers and men assigned to all regiments. These men are distributed so that there is a medical man with each platoon as well as a small medical section with the regiment. The duties of the platoon medical men include preliminary treatment of the wounded, in most cases amounting to the application of the first field dressing carried by every man, and the continuation of treatment of cases returned from hospital. It is also apparent that they are to ensure that very strict prophylactic discipline is observed to counter epidemics and other avoidable diseases.

Compared with other military organizations around the world, the Japanese assigned a far greater number of army personnel to infantry than was the normal course. Army policy was to minimize logistics in order to place the maximum number of troops on the front lines. As a result, shortages of food, medical supplies, and even ammunition were commonplace.

By 1943 American commanders in the Pacific had become deeply suspicious of the heavy hospital ship traffic steaming under the Japanese flag. That suspicion that was confirmed by American intelligence which decoded Japanese messages showing the Japanese were attempting to use the hospital ships to transport troops. Because of the great secrecy surrounding cryptanalysis meant that the Allied commanders could neither act on nor publicly disclose these violations.

Malnutrition

Victims of malnutrition included not only POWs held by the Japanese, but also the Japanese garrisons bypassed by the Allied counteroffensive, such as those manned by the Seventeenth and Eighteenth Armies. Shortages of vitamin A in the diet led to night blindness, while shortages of vitamin C caused scurvy. Scurvy is characterized by anemia and spongy gums. Deficiency of thiamine (Vitamin B1) caused beriberi, which was characterized by neuropathy and damage to the heart. Lack of niacin (Vitamin B3) produced pellagra, characterized by dermatitis, diarrhea, and dementia. Almost all vitamin deficiencies slowed the healing of even minor wounds. The only treatment was administration of the missing vitamins. Japanese troops sometimes showed signs of beriberi even in garrison. Any breakdown in logistics rapidly caused large numbers of Japanese troops to come down with the full-blown illness. This was evident at Guadalcanal, where large numbers of Japanese were encountered during their retreat who continued to wield a rifle even though they could no longer stand due to neuropathy in their lower limbs.

The Japanese Army tended to supply food in bulk and often raw and uncooked. There were no individualized portions, such as the well-known GI C-rations. Japanese food usually was delivered in 50-gallon drums or in bushel sacks of rice and barley. Some drums were air dropped while others arrived by sea on surface or submarine vessels. Commonly, individual soldiers received A-rations, two-day supplies of uncooked polished rice, some hard-tack, perhaps a tin of meat, and a fuel packed in a bamboo tube. It was then up to the individual soldier to cook and prepare his own meals. When barley was unavailable the Army frequently substituted a bean paste. It also occasionally supplied soy sauce, corn yeast, dried plums, dehydrated egg yolk, salt fish, and pickled radishes. Supplies arrived sporadically and the daily ration of 870 grams of rice was frequently reduced by one-half to two-thirds.

Soldiers were expected to forage for supplements to the meager daily rations. On many Pacific islands that was impossible because of the lack of edible plants. Additionally, Japanese soldiers had no way of knowing which plants, or parts thereof, might be edible, especially if properly harvested and prepared. The Army had not prepared guides to choosing local plants. Garrison troops, and those deployed on long term occupancy, were encouraged to grow their own food. This met with extremely mixed results. Troops on New Guinea especially complained about the lack of useful native plants and the miserable growing conditions for native Japanese plants. Troops on islands where coconuts grew rejoined in harvesting their meat and milk when in season.

Beriberi

Beriberi is a disease of the nervous system caused by a person not getting enough thiamine (vitamin B1) in the diet. Thiamine is needed to break down food such as glucose. It is also found on the membranes of neurons. Symptoms of beriberi include severe lethargy and tiredness. There are two kinds of beriberi, wet and dry. Wet beriberi mainly affects the cardiovascular system. Dry beriberi affects the nervous system. There is also a rare type of genetic beriberi. This disease is found in Asia. There, some people only eat white rice. This disease can make it hard for a person to do simple things. It can also cause muscles to get weaker. A person with beriberi may also have heart problems or even death. Frequently, dry beriberi and wet beriberi occur at the same time in the same person. When this takes place, the victim is likely to only get the symptoms of one form of beriberi.[289] From the beginning of the Japanese medical units, the great concern of the Japanese physicians was the incidence of beriberi, known in Japan as Kak'ke. It was initially though that beriberi was a disease that was communicated from an infected person to a healthy one. That explanation was quickly rejected. It caused over half of the sickness in the military before 1900.

---

[289]"Beriberi" Wikipedia; Cambridge English Dictionary.

Baron Kanehiro Takaki undertook a study that concluded that processed white rice, a staple of both the Imperial Army and the Imperial Navy, was to blame.[290] Takaki discovered that beriberi stems from a lack of vitamin B1, which the body requires for metabolizing carbohydrates and maintaining neurological functions. Without it, a person succumbs to nerve damage and eventually death. Adding barely, or at least rice husks, to the diet reduced, then eliminated, the disease. Feeding chickens a diet of polished rice quickly brought about conditions similar to beriberi in humans. Returning to a more balanced diet quickly restored the chickens to health. Sufferers in the Army frequently reported ulcers and a prickly hot feeling on the bottoms of their feet. Many upper-class Japanese, unwilling to abandon white rice and fish diets began to inject themselves with vitamin B to prevent beriberi.

By correcting the vitamin deficiency by augmenting the rice and fish diet the Japanese navy gained a major victory in the Russo-Japanese War of 1904. It was common for all armies of the world, through the end of the 19th century, to lose more men to disease than to war wounds. Twenty years after Takaki had eradicated beriberi from the navy, Japan faced Russia in the Russo-Japanese War from 1904 to 1905. The army, which had ignored Takaki's findings, hospitalized 250,000 soldiers with beriberi, 27,000 of whom died. The shocking toll ended the army's resistance. In the middle of the war in February 1905, Gen. Masatake Terauchi ordered the army to mix barley with its rice.[291] Concern for the dietary imbalance remained a priority for the Japanese military through the Pacific War.

While commanded to forage and live off the land as much as possible, Japanese soldiers assigned to most Pacific islands found that impossible. Many islands had few, if any, edible plants. There was rarely

---

[290]Basil Hall Chamberlain, a preeminent Japanologist, demonstrated the prevalent lack of understanding of the disease's causes in *Things Japanese: Being Notes on Various Subjects Connected with Japan* (1890). The disease springs, in the opinion of some medical authorities, not from actual malaria, as was formerly imagined, but from a climatic influence resembling malaria . . . . "Others have sought its origin in the national diet—some in rice, some in fish."

[291]Seaman, Louis Livingston. *The Real Triumph of Japan*. New York: Appleton, 1907, pp. 120ff.

a surplus, a supply beyond the basic needs of natives. Beriberi appeared because of the lack of vitamin B. Japanese records show that beriberi appeared in both the Philippines and in Singapore. The army led by Major-general Horii which was supposed to trek over the Owen Stanley Mountains and attack Port Mores by was rendered impotent by beriberi and general malnutrition. Fully one-fourth of the men who landed with Horii were suffering from disease within three weeks of landing and about one-half were afflicted within five weeks.[292]

Throughout the war, beriberi remained a significant problem for the Japanese Army. And there were other related illnesses. Night blindness, which results from a deficiency of vitamin A, afflicted many Japanese soldiers. Scurvy ran rampant in some infantry units, a result of the lack of vitamin A. Even the best supplied infantry units suffered some cases of vitamin deficiencies. The high incidence of beriberi can be cited as one of the most blatant signs of Japanese bureaucratic neglect of its Japanese soldiers.[293]

Malaria

Malaria was among the greatest problem for the Japanese medical services. This disease is transmitted by mosquitoes. This disease, while not typically fatal to the infected soldier, would take him out of action for a prolonged period just as surely as if he had been wounded in battle. Malaria was at least as serious a problem for the Japanese Army as it was for the Western Allies.[294] At the end of the Bataan campaign in 1942, over seventy percent of one unit of the Imperial Japanese Army had malaria. Two years later in Montana, sixty percent of the troops had malaria, of

---

[292]Gailey, Harry. *MacArthur Strikes Back: Decision at Buna, New Guinea 1942-1943*. Novato, CA: Presidio, 2000.

[293]Lock, M. *Asian Medicine in Urban Japan*. Berkeley: University of California Press, 1980.

[294]About 24,000 of the 75,000 American and Filipino soldiers involved in the desperate campaign to stop the Japanese advance in 1942 were suffering from malaria. As malaria affected Allied troops during the Marshall and Solomon island campaigns, Lieutenant General Robert L. Eichelberger, commander of the U.S. forces bogged down at Buna, noted that "disease was a surer and more deadly peril to us than enemy marksmanship. We had to whip the Japanese before the malarial mosquito whipped us."

which twenty percent were hospitalized and ten percent soon contracted colitis, which usually resulted in death. After 45 days of medical treatment most patients were able to return to their units. To an effort to prevent malaria, Japanese troops were issued mosquito repellent, one type of which came in a green glass bottle, and another came in an olive drab painted tin. A Japanese physician noted that it smelled like a common Japanese perfume. Yet another repellent was in the form of a spiral shaped incense stick that could be burned to drive away mosquitoes.

The Imperial Japanese Army used mosquito nets large enough for a whole squad. Medics made anti-mosquito cream and spray available to sentries and others who had to be outside the net. They also administered Brattain or quinine every day to all who had not contracted malaria. In spite of these precautions nearly every man in most Japanese infantry regiments had contracted malaria at least once during his service.

Apart from malaria the Japanese took great care to prevent epidemics that might spring from impure drinking water. The platoon medical orderly was held responsible for implementing water hygiene. When the army medical authorities were not been able to procure a supply of pure water, they practiced water purification by employing chlorinating vials which were carried by every man in a platoon.

In addition to anti-malaria and water precautions, the medics issued various pills to every man however healthy. One pill was taken every ten days containing vitamin B. Other pills were taken every day, or every five days, presumably to maintain general health. The Japanese were also strict about mandating inoculations and vaccinations. Battalions were often inoculated en masse.[295]

At the beginning of World War II, the world supply of quinine was largely under control of the Netherlands. The island of Java, a Dutch colony, was the world's largest producer of cinchona, which thrived in its tropical climate. A consortium of growers banded together to form the Kina Bureau, based in Amsterdam, which dictated the supply of quinine and held an effective monopoly on this valuable commodity. When Germany occupied the Netherlands in May 1940, it gained access to the

[295]Sharpe, George. *Brothers Beyond Blood: A Battalion Surgeon in the South Pacific.* Austin: Diamond Books, 1989.

warehouses and manufacturing capabilities of the Kina Bureau. When the Japanese occupied Java, along with the rest of the Dutch East Indies in 1941, they controlled the raw supply of cinchona. As the war progressed, the Allies were increasingly effective in cutting off Japanese access to various materials that were plentiful in areas it had conquered, and that included quinine. Japanese troops soon were suffering extreme rates of malarial infection. It was estimated that at times, up to ninety percent of Japanese troops in some locations were combat ineffective due to malaria and dysentery. After the Battle of Buna, the U. S. Third Portable Surgical Hospital records noted that every single Japanese prisoner they received was infected with malaria. Thus, Japan had all the quinine its soldiers required, but the problem was, as with so many things for Japan, the logistics of distribution.[296]

In addition to, and caused by the same conditions that fostered malaria, other tropical ailments such as ringworm, fungal infections, dysentery, and beriberi were nearly ubiquitous among the men serving in jungle environments. All these diseases had a strong effect on the soldiers of all nations which were engaged in the Pacific War.[297]

Japanese Medics

The Japanese military did have an established medical corps with echelons of care. At the platoon level minimally trained soldiers acted as medics in the field. A typical medical orderly attached to a platoon carried a soldiers common knapsack containing the following: peptic tablets for stomach trouble; adhesive plaster; aspirin tablets; bandages; a morphine solution; gauze; tincture of iodine; scissors; iodoform which was an antiseptic; a thermometer; zinc oxide; boric acid; atabrin rivanol solution which was a disinfectant; quinine sulphate; a syringe; sodium bicarbonate; a sharp knife; and absorbent cotton. The most forward army medical unit for the treatment and evacuation of casualties is the advance dressing station of the divisional medical unit. There the patients are classified and

---

[296] Paltzer, Seth. "The Other Foe: The U.S. Army's Fight against Malaria in the Pacific Theater, 1942-45" U.S. Army National Museum, 30 April 2016.

[297] Gillespie, James O. "Malaria and the Defense of Bataan." Army Medical Department, Office of Medical History. http://history.amedd.army.mil/ booksdocs/ wwii/Malaria/chapterIX.htm

their preliminary dressings checked. They are then evacuated by means of the ambulance company of the same unit to one of the division field hospitals. There was usually one of these with each first line regiment and they are equipped to perform reasonably extensive surgical treatment.

There were battalion-level aid stations which were largely staffed by physicians. Battalion dressing stations [Sento Kyugo Han] would have been staffed with two or three medical officers and up to ten enlisted men. These facilities were to provide emergency first aid and care for the sick. The officers were sometimes asked to present lectures on disease prevention and hygiene to assembled companies and platoons. About 1943, the Army made changes in its medical department. The battalion dressing stations were combined with the Field Hospitals in order to simplify the medical evacuation route.

At full strength, a division medical unit [Eisei Tai] employed over 550 men, organized into three companies. Two companies evacuated the wounded. A casualty collecting company provided litter bearers to the clearing stations and field hospitals. A casualty clearing company evacuated patients from field hospitals to various communication hospitals. These units were rarely equipped with ambulances but rather used trucks, boats, native carriers and various beasts of burden, such as elephants in Burma. The division water supply and purification company provided sanitation.

At the next level were regimental hospitals staffed with skilled specialists. The Division Surgeon controlled four field hospitals, one for each regiment. These hospitals could be assigned according to need and split up, if necessary. Each hospital could accommodate 500 patients and was equipped to perform surgery and do blood transfusions. Patients were placed in straw bedding because field hospitals were not outfitted with beds. Finally, some wounded may have been scheduled for evacuation from theater. From the division field hospitals patients are further evacuated to line of communication hospitals, reserve hospitals or base hospitals in the home areas or are released to rejoin their units. Evacuation is performed with a combination of stretcher, horse-cart ambulance, motor ambulance, ambulance trains, river craft and by hospital ships.[298]

[298]*Notes on Japanese Medical Services. WWII* Tactical and Technical Trends, No. 36,

Line of communication hospitals [Heitan Byoin] were commonly located approximately fifteen miles to the rear of the field hospitals. Some were large enough to accommodate 1,000 patients. Those taken to these hospitals could be evacuated to the Japanese mainland. These hospitals were staffed with as many as twenty-five medical officers, two or three dental officers and ten nurses. They had units created for the treatment of gunshot fractures and other specialized care units. A mobile treatment section could be detached to administer first aid wherever necessary.

One rarely hears hear much about the Japanese care of their wounded. First, the Japanese military in active combat against the Pacific suffered appalling losses, often fighting to the point of total destruction. Correspondingly outside of very rare diaries, or the few POWs, many battles are only known extensively through the Allied accounts, or the high-tier Japanese logs.

Second, the capabilities of Japanese medical attention was very poor compared to their allied foes. The U.S. medical establishment especially was at the end of a massive logistical effort, with an extensive system of out of theater evacuation for the severely wounded. The Japanese generally had to make due with much less, and usually were fighting isolated from their supply lines due to successful sea interdiction.[299]

Supplies of morphine were sometimes exhausted by heavy casualties when the IJA engaged Western forces, forcing physicians to reserve the supply for the worst casualties. However, accounts of surgery without anesthesia are so numerous from Japanese veterans that they suggest a cavalier attitude towards the suffering of wounded soldiers and sailors.

The failure of the Japanese to respect the Red Cross convention led the Allies to give their medics "defensive arms," which the Allies claimed were not prohibited under the Conventions if used only to protect medics and their patients. Frequently, medics removed the Red Cross

October 21, 1943.At full strength, a division medical unit [Eisei Tai] employed over 550 men, or more.

[299]Toland, John. The Rising Sun. This book includes an extensive account of Japanese medics written by a Japanese civilian pressed into augmenting a military field hospital staff.

brassard. Bandages were dyed green because Japanese sharpshooters spotted white bandages. Allied intelligence claimed in 1944 that Japanese medics routinely carried arms and frequently were used as replacements in combat units. This was common when the Japanese abandoned any hope of victory and decided to make a final banzai charge.

Fact and Fiction

What the U.S. Army took to be the way Japanese troops would be cared for when injured or wounded, was far more fiction than fact. In general, the Japanese Army did not have field kitchens, field hospitals or a dedicated medical corps. Over 60% of the deaths of their soldiers was the result of starvation and disease. In fact, it is surprising that, following years of blunders and disasters, the Japanese officers weren't killed by their own troops for gross incompetence.

When the Japanese Army had to retreat, the wounded were usually left behind. All Japanese soldiers were acutely aware that the medics and physicians had been instructed to either shoot their patients, or give them a hand grenade and encourage them to kill themselves. That flowed from the fundamental assumption that any form of surrender was unacceptable in the eyes of the Japanese. Medics rarely had access to any sort of transport for the wounded. It was even more acute for those who wounds were so serious that they could not move themselves and for those whose legs had been amputated.

It was common for Japanese physicians and especially medics to run out of vital medical supplies if an engagement took longer than expected. During the Battle of Shanghai in 1937, the battle continued for nearly three months before the city was secure. Some Japanese soldiers who had contracted diseases like dysentery were herded into open pens and cordoned off to establish a makeshift quarantine. Naturally, these conditions greatly worsened the suffering of the men in question.[300]

Meth

A Japanese chemist first synthesized methamphetamine in 1893.

[300]Harries, Meirion. *Soldiers of the Sun: The Rise and Fall of the Imperial Japanese Army. Random House,* 1991.

Meth, as it is commonly known, was first developed as a man-made alternative to the ephedra plant. Ephedra is a type of shrub whose extract has been used in traditional Chinese medicine for over 5,000 years. In 1885, Nagai Nagayoshi, a Japanese chemist studying in Germany, identified the active chemical in ephedra, a stimulant called ephedrine. Methamphetamine was difficult to make until 1919, when Akira Ogata, a second Japanese scientist, streamlined the process. He used phosphorus and iodine to reduce the ephedrine into a crystallized form, creating the world's first crystal meth. Methamphetamine was used early on as a medical treatment for narcolepsy, asthma and as a weight-loss drug. During World War II, Japan used the drug to keep troops awake.[301] In 1954 Japanese psychiatrists told of the high incidence of amphetamine that were used extensively during World War II to stimulate civilians as well as soldiers to greater effort. Amphetamines remained readily available after the war so that there were at least 200,000 amphetamine addicts in Japan by the 1950s. The psychiatrists studied 492 Japanese methamphetamine addicts and found that chronic psychoses resulted in many of these addicts, some of whom have been permanently affected, with experimental and clinical evidence of organic brain damage.[302]

---

[301]Editors of History.com. "History of Meth" Originally posted on 7 June 2017 and updated on 21 August 2018. https://www.history.com/topics/crime/history-of-meth

[302]Lemere, Frederick. "Amphetamine Addiction in Japan" *JAMA*. 1963; 185 (5): 414.

Comments

The photographs that appear sporadically and also at the end are public domain, removed from various places on the world wide net. None are pretty to look at; indeed, they run the course from total disgust to evocation of sorrow to evoking great anger. They are absolutely authentic. No photo-shop fabrications here.

How is it that a nation can perpetrate such horror on fellow human beings? How can a nationality consider itself so wholly superior to other nations that they have some sort of natural right to treat the others as inferiors? How can individual human beings, unless insane, do such things to other human beings?

If we could answer these questions, we would be geniuses. We would have solved the great mystery of the twentieth century. So far no one really has come up with a reasonable explanation for human conduct that violates every norm of humanity.

Ever since the defeat of the Third Reich, Germany has been making atonement for the horrors of Nazism. It has truly come face to face with its past. In case it falters in any way there are those who will remind the German people of their past. We are now past the time when any remaining survivors of Hitler's willing executioners can be brought before a court, since age has overtaken those few. Those of us who believe in the Final Judgment when all humans will stand before the Throne of God and be forced to admit our sins, we believe that there the guilty will perish in pain from that Final Judgment.

But, as a nation, Japan has not acknowledged its guilt. Japanese textbooks for the public-school system rarely mention any details of the Pacific War. If and when that war is a subject it is because the writers wish to draw attention to American use of the atomic bomb. Perhaps the fire-bombing of Japanese cities will draw mention. But, one can bet, there will be no mention at all of the subjects mentioned in this booklet. Japan does not acknowledge the inhumane treatment of prisoners of war. There will be no notation of the Japanese colonel who killed an American flier and ate his liver. Nothing is said of the occupation and incredibly barbarous treatment of Korea or Manchuria or China. There is no chapter on enforced prostitution of the so-called comfort girls in the Philippines or China.

Although General Douglas MacArthur allowed no photographs of the executed war criminals, starting with Hideki Tojo, it is almost certainly true that some Japanese managed to salvage the ashes of those cremated bodies to enshrine them. Tojo's commemorating tomb is located in a shrine in Nishio, Aichi, Japan, and he is one of those enshrined at the controversial Yasukuni Shrine. His ashes are divided between Yasukuni Shrine and Zōshigaya Cemetery in Toshima ward, Tokyo. Somehow, with or without General MacArthur's knowledge. there had been an enshrinement of the ashes of fourteen Class-A war criminals in Yasukuni Shrine. They are commonly known as martyrs. This stands in bitter contrast to the fate of the Nazi war criminals whose ashes were discarded in a stream so that none could ever be so enshrined.

Tojo's granddaughter, Yūko Tojo, was a far-right ultra-nationalist and political hopeful who gained both some interest and following in post-war Japan. She claimed that the Pacific War was one of self-defense. Like her grandfather, she defended war generally, looking on it as something nearly sacred, with a legitimate purpose in the world. Her grandfather hallowed that war by his actions and by his ultimate sacrifice. She lectured widely on the theme that it was unfair that her grandfather was judged a war criminal and executed. He was, in her view, a true patriot who justifiably served his native country.

Most modern historians, such as Herbert P. Bic and John W. Dower, have criticized the United States for protecting the emperor. In 1945 Americans believed President Harry Truman that Japan had surrendered unconditionally, as required by the Potsdam Declaration of the great powers. This was simply not true because Truman had accepted the Japanese condition that the emperor be allowed to remain on his throne. After all, he was a demi-god, accepted as such by the vast and overwhelming majority of Japanese.

Protection of the emperor required much work to be done by

General MacArthur and his staff. They had to find some reasons to exonerate Emperor Hirohito (1901-1989), the 124th *mikado,* and all members of the imperial family from criminal prosecutions. General MacArthur and Brigadier General Bonner Fellers worked to protect the Emperor by shifting the ultimate responsibility to Tojo. According to reports, General Fellers met with various Japanese at his office and told them that it was highly important that the Japanese prove that the Emperor was completely blameless. Tojo, in particular, should be made to bear all responsibility at this trial. As a most loyal Japanese with deep regard for the emperor Tojo did exactly that.

But this was a great lie, a conspiracy without merit. To claim that the military had acted without the connivance of the emperor was unthinkable. There was, indeed, a brief time in Japanese history when the *mikado* was purely a figure-head, a quasi-religious figure, but without political power, that time was not during the reign of Hirohito. For example, Admiral Nagano had explained to the Emperor in detail the plan of the attack on Pearl Harbor. The fact that Hirohito had intervened and mandated Japan's surrender told something of his real power.

Moreover, there were the diaries of Marquis Kōichi Kido, Lord Keeper of the Privy, longtime adviser to the emperor. The chief prosecutor, Joseph B. Keenan of the United States, an alcoholic appointed by President Harry S. Truman, used Kido's diaries as a road map for the prosecution. Kido's diaries showed that Hirohito was anything but a figure-head who presided but did not rule. He was consulted at every turn and in every decision. His concurrence was absolutely required for every move.

The sustained intensity of this campaign to protect the Emperor was revealed when, in testifying before the tribunal on December 31, 1947, Tojo momentarily strayed from the agreed-upon line concerning imperial innocence and referred to the Emperor's ultimate authority. The American-led prosecution immediately agreed to an adjournment during which time Tojo was secretly coached on how to recant this testimony. Keenan used Ryūkichi Tanaka, a former general who testified at the trial, to bring Tojo back to the line both defense and prosecution had agreed upon.

Not only did General MacArthur exonerate the Emperor, but also all members of the imperial family who were implicated in the war. The protected figures included Prince Chichibu, Prince Asaka, Prince Higashikuni and Prince Hiroyasu Fushimi. Chichibu was one of the members of the Imperial household who was personally involved in stealing treasures from countries invaded by Japan during World War II. One fact revealed in Kido's notes was that the uncle of the emperor had been present at the Rape of Nanking and, if not actually ordering it, had concurred completely. But, for the whole royal family, there would be no criminal prosecutions.

Hirohito was turned into an almost saintly figure who did not even bear moral responsibility for the war. Not only did Hirohito survive the trial completely unscathed, he lived to become an internationally known figure. Hirohito became an active figure in Japanese life. He performed many of the duties commonly associated with a constitutional head of state. He and his family maintained a strong public presence, often making public appearances at special events and ceremonies. He also played an important role in rebuilding Japan's diplomatic image. He traveled abroad and met with many foreign leaders, such as British Queen Elizabeth II in 1971) and U. S. President Gerald Ford in 1975.

The worst criminals to escape justice were the operatives of Unit 100 and Unit 731. Several of the officers had flown out of Manchukuo just ahead of the invading Soviets. In later life they became millionaires in the sale of human blood on the commercial market. There was no one to prosecute them. The Japanese were wholly unconcerned about these men without consciences.

Germany has captured, or had extradited, a number of individuals accused of various war crimes. Japan has never sought any of its unindicted war criminals. Many of the most brutal guards at POW Camps escaped human justice. No Japanese governmental agency was established to look into war crimes of any kind.

Japan escaped into the modern world where its industrial capacity has gained far more economic advantages than ever did their conquests. But it escaped without a conscience-driven desire to atone for its brutal past.

The Japanese received an unexpected boost from the Tokyo Trials in the form of the strong dissenting opinion written by Indian Justice Pal. Initially suppressed, the Pal dissent became a basis of a Japanese court case seeking to condemn American use of atomic bombs. It also is used to justify the Japanese refusal to confront war crimes.

Radhabinod Pal (1886 –1967) was an Indian jurist, who was a member of the International Law Commission of the United Nations from 1952 to 1966. He was one of two Asian judges appointed to the International Military Tribunal for the Far East, known as the "Tokyo Trials," of Japanese war crimes committed during the Second World War. He was the only one of the judges who submitted a dissenting opinion. Pal insisted all defendants should have been found not guilty. The Yasukuni Shrine and the Kyoto Ryozen Gokoku Shrine have monuments specially dedicated to Judge Pal. Radhabinod Pal was a judge chosen to represent the interests of India, a nation which had long been under British colonial rule. Pal was strongly critical of Western colonialism, while, on the other hand, he naturally had a deep sympathy with what he considered to be the struggle of modern Japan against political and economic oppression imposed by the Western nations. Publication of Pal's dissenting opinion at the Tribunal was prohibited during the Occupation years, and even after the Allied Occupation was over and Japan had regained its independence. Finally, it was published by the Kenkyusha Publishing Company, Tokyo, in November 1976, as a textbook for college use, under the title *In Defense of Japan's Case*.

Pal was far more critical of the West for using the atomic bomb than he was of Japan for its aggressive war. He lauded the emperor because Pal believed that it was only because of his caution that Tojo abandoned Japan's atomic bomb program. Otherwise, Pal wrote, Japan might have won the war.

To the Japanese, Pal's opinion was deeply moving partially because it was based upon an Asian understanding of Asian history, but largely because it seemed to accept the explanations of the Japanese defendants. Pal argued that the events that led up to the war could be explained by means other than an aggressive conspiracy, Pal devoted the greater part of his judgment to the historical background of disputes, incidents and wars. Indeed, his opinion reads more like a lecture on Asian history, rather than a legal opinion.

Justice Pal complained bitterly about the exclusion of Western colonialism and the American use of the atom bomb from the list of crimes. He also objected to the exclusion of judges from the vanquished nations on the tribunal. As long as major questions and charges were omitted the tribunal could not claim a moral existence.

War in Asia, Judge Pal argued, is fought by standards that are wholly different from, and incomprehensible to, Western nations. War has a kind of natural and legitimate existence, and during its waging a set of moral standards that differ substantially from peaceful norms come into existence. Pal never challenged the evidence that atrocities were perpetrated by the members of the Japanese armed forces against the civilian population of some of the territories occupied by them as also against the prisoners of war, there was no grounds for prosecuting the leaders after the war was over. Certain horrific events, such as the Nanking Massacre, could be settled by lower level prosecutions.

Pal advanced the view that the legitimacy of the tribunal was suspect and questionable, because the spirit of retribution, and not impartial justice, was the underlying criterion for passing the judgment. Since the trial was held with evil intent [*mens rea*] its verdicts could not be sustained in the court of human ethical conduct. Judge Pal never intended to offer a juridical argument on whether a sentence of not guilty would have been a correct one. However, he argued that the United States had clearly provoked the war with Japan and expected Japan to act. In his dissent, Judge Pal referred to the trial as a "sham employment of legal process for the satisfaction of a thirst for revenge." If the trial was illegitimate so must its verdict be extra-legal and therefore rejected.

Post-war Legal Activities

There were two interesting post -war developments, one each in American and Japanese courts, whose ultimate decisions buttressed one another. Former U.S. prisoners of war attempted to sue Japanese industrialists for employing POWs as forced labor assisting the Japanese war effort while Japanese citizens, survivors of atomic bombing, attempted to sue the United States for violations of various laws dealing with land warfare. Neither case succeeded although the Japanese case may be regarded more as a propaganda move aimed at discrediting the United States than as a real attempt to recover damages.[303]

Beginning immediately after the end of the Pacific War Japan tried to rewrite history. Within days of Japan's surrender on August 15, 1945, coded messages went out from Tokyo to Japanese diplomats which ordered them to undertake a campaign which portrayed Japan as a victim in the war. Most importantly, the government ordered all personnel to play down Japanese atrocities while emphasizing Japanese civilian losses from bombing, and especially casualties in the recent atomic bomb attacks. These particular messages were not decrypted by the United States until years after the war because there were other priorities so these Japanese the messages were simply recorded and filed away. By the time the Japanese messages were deciphered, the Cold War had begun, and the West needed Japan as an ally against the communist advances. Those decrypted messages were then kept secret for decades, along with most of

---

[303]. There is a recent Japanese book of interest. Tanaka, Yuki. *Hidden Horrors: Japanese War Crimes in World War II.* Westview, 1998. This book is a discussion of Japanese war crimes from a Japanese perspective, with the author's theories on why such crimes took place. This should make it a valuable addition to the literature. However, the author dilutes both individual and cultural responsibility in his argument that Emperor-worship in the Japanese military was the primary influence bringing out the universal human propensity for evil in the particular individuals responsible for war crimes.

the details of how Allied code breakers had read most of the enemies most secret messages throughout the war. What was not so secret were Japanese efforts to ignore the war and portray themselves as victims. Few Japanese opposed this blatant rewriting history, which was often quite creative. While this tampering with historical facts caused problems with Japan's neighbors, especially China, the Cold War prevented the West from taking any action. Thus, the Japanese were generally most successful in evading virtually any responsibility.[304]

The Shimoda Case

Beginning immediately after the atomic bomb attacks on Hiroshima and Nagasaki, there has been legal debate over the legality of the action. On 10 August 1945, the Japanese government addressed a communication to the International Committee of the Red Cross, asking it to denounce the U.S. government for commission of two crimes under international law. Following surrender and the landing of U.S. occupation troops in Japan, the Prime Minister Naruhiko Higashikuni offered not to file complaints in the media or take legal actions because of the use of the nuclear weapons provided the United States agreed to abandon its plan to try Japanese war criminals. The U.S. rejected that offer and General Douglas MacArthur proceeded with plan for the Tokyo War Crimes Trials.

Once the War Crimes Trials opened, some of the defense lawyers tried to persuade the International Military Tribunal of the Far East to include charges against the United States relative to the matter of the legality of the first use of nuclear weapons. These motions were ignored as the tribunal had long since precluded any defense of tu quoque or of allowing inquires into Allied conduct of the Pacific War. One defense lawyer, Shoichi Okamoto, continued to pursue the issue with great vigor long after the trial had been concluded. In February 1953, he published a booklet titled *Questions and Answers on the Civil Lawsuit over the Atomic Bombings*. In that inflammatory monograph he sought individuals in Hiroshima and Nagasaki who would be willing to take legal action against

[304] Dunnigan, James. "Japan Wants to Rewrite World War II." *Strategy Page*, 22 August 2010. https:// www.strategypage.com / dls / articles / Japan-Wants-To-Rewrite-World-War-II--8-22-2010.asp

the United States, but within the U.S. court system.

The Japanese themselves rejected Okamoto's plan. That scheme met great deal of opposition within Japanese society, even in Hiroshima and Nagasaki, and for a variety of reasons. For one reason, the mayor of Hiroshima opposed the plan, arguing that the U.S. legal system was not favorably disposed to accept to such actions. For another, there were realists who recognized that Japan had irrevocably waived access to U.S. courts in its surrender and subsequent Treaty of San Francisco.

Therefore, Okamoto jettisoned the notion of trying the case in a U.S. court and decided to seek action in the Japanese legal system. He and his cooperating legal and political team selected a group of five people who would become the litigants in a Japanese court. Another lawyer, Yasuhiro Matsui, then joined the legal team. Shimoda, the leader of the group, came from Hiroshima and was 57 years old. He lost four daughters and one son in the atomic attack on Hiroshima, and he, his wife and surviving son suffered from persistent health problems.

The case of *Ryuichi Shimoda et al. v. The State*[305] involved unsuccessful litigation brought before the District Court of Tokyo by a group of five survivors of the atomic attacks on Hiroshima and Nagasaki. These litigants claimed the dropping of the atomic bomb was illegal under the laws of warfare as they existed in 1945. They sought reparations from the Japanese government on the ground that it had waived the right to sue for reparations from the U.S. government once it had signed the Treaty of San Francisco. Specifically, the litigants cited the Hague Convention of 1907, Title IV, The Laws and Customs of War on Land; and Title IX, Bombardment by Naval Forces in Time of War; and also the Hague Draft Rules of Air Warfare of 1922–1923.

The plaintiffs were all Japanese nationals, and, at the time the atomic bombs were dropped, residents either of Hiroshima or of Nagasaki. There was no question but that the atomic bombs were dropped on these

---

305 *Shimoda et al. v. The State*, the Wayback Machine, Tokyo District Court, 7 December 1963. ICRC, International Humanitarian Law, National Implementation: National Case Law: *Shimoda* case. See also http://www.weapons law.org/cases/japan-tokyo-district-court-shimoda-1963

cities by bombers of the United States Army Air Corps in August 1945. The result of the bombing was that most of the members of the plaintiffs' families were killed and many, including some of the plaintiffs themselves, were seriously wounded as a result of these bombings. They alleged that the aerial bombardment with the use of atomic bombs of the cities of Hiroshima and Nagasaki was an illegal act of hostilities according to the rules of international law.[306]

The plaintiffs as individuals were not entitled to claim damages on the level of international law, and were not able, as a result of the doctrine of sovereign immunity, to pursue a claim on the plane of municipal law, that the waiver of claims in the San Francisco peace treaty precluded any actions for damages by Japanese citizens against the US government, was also used in the U.S. legal system. Thus, the plaintiffs had no rights to lose as a result of the waiver contained in Article 19 (a) of the San Francisco Treaty of Peace with Japan.[307]

The plaintiffs sought monetary damages, alleging that

> (a) that they suffered injury through the dropping of atomic bombs by members of the [Army] Air Force of the United States of America; (b) that the dropping of atomic bombs as an act of hostilities was illegal under the rules of positive international law (taking both treaty law and customary law into consideration) then in force, for which the plaintiffs had a claim for damages; (c) that the dropping of atomic bombs also constituted a wrongful act on the plane of municipal law, ascribable to the United States and its President, Mr. Harry S. Truman; (d) that Japan had waived, by virtue of the provisions of Article 19 (a) of the Treaty of Peace with Japan of 1951, the claims of the plaintiffs under international law and municipal law, with the result that the plaintiffs had lost their claims for damages against the United States and its President; and (e) that this waiver of the plaintiffs' claims by the defendant, the State, gave rise to an obligation on the part of the defendant to pay

[306] Yasuhiro, Matsui. *The A-Bomb Trial: Legal Principles for Abolishing Nuclear Weapons and Supporting A-bomb Survivors.* Shin-Nippon Shuppan-sha, 1986.

[307] Tanaka, Yuki and Falk, Richard. " The Atomic Bombing, The Tokyo War Crimes Tribunal and the Shimoda Case: Lessons for Anti-Nuclear Legal Movements" in *Asia-Pacific Journal*, https://historynewsnetwork.org/ article/119579. Richard Falk's article was reproduced in his book, *Legal Order in A Violent World.* Princeton University Press, 1968, pp.374 – 413.

damages to the plaintiffs.[308]

More specifically, the plaintiffs based their case on the provisions of Article I of the State Redress Law, which was applicable to the case of injury to a private person through an unlawful act of a government official; on the provisions of Article 29 of the Constitution, which provided for the obligation to pay just compensation in every case of expropriation of private property by the State for public See also use; and, finally, on unlawful infringement of the rights of the plaintiffs through the omission of the defendant to take appropriate measures for recovery of compensation.

Proceedings at the District Court in Tokyo began in April 1955, and they lasted for eight and a half years.[309] The final ruling was rendered was not rendered until December 8, 1963.[310] Okamoto, who died of a stroke in April 1958, did not live to see the final ruling. On the twenty-second anniversary of the Japanese sneak attack on Pearl Harbor, the District Court of Tokyo declined to rule directly upon the legality of nuclear weapons in general. However, it did find that "the attacks upon Hiroshima and Nagasaki caused such severe and indiscriminate suffering that they did violate the most basic legal principles governing the conduct of war." In the opinion of the court, the act of dropping an atomic bomb on cities was at the time governed by international law found in, and was therefore illegal.[311]

The Court drew a distinction between "Targeted Aerial Bombardment" and indiscriminate area bombardment, that the court called "Blind Aerial Bombardment" and a distinction between a defended and an undefended city. "In principle, a defended city is a city which resists an attempt at occupation by land forces. A city even with defense installations and armed forces cannot be said to be a defended city if it is far away from the battlefield and is not in immediate danger of occupation by the enemy." The Japanese court held that the aerial bombardment with

---

[308]It was reported in the *Hanrei Jiho,* vol. 355, p. 17; translated in T*he Japanese Annual of International Law*, 8: 231 (1964).

[309]Argued April 1955-October 1963, Decided December 7, 1963.

[310] *Japanese Annual of International Law*, 8: 212 (1964).

[311]Falk, Richard and Mendlovitz, S. eds., *Towards A Theory of War Prevention*. World Law Fund, 1966.

atomic bombs of the cities of Hiroshima and Nagasaki was an illegal act under the rules of international law as they existed in 1945.[312]

The court ruled that blind aerial bombardment is permitted only in the immediate vicinity of the operations of land forces and that only targeted aerial bombardment of military installations is permitted further from the front. It also ruled that the incidental death of civilians and the destruction of civilian property during targeted aerial bombardment was not unlawful. The court acknowledged that the concept of a military objective was enlarged under conditions of total war, but stated that the distinction between the two did not disappear. The court regarded the attacks as indiscriminate aerial bombardment of undefended cities in clear violation of the law of nations. Even if the attacks had been directed against only military objectives, such bombardment resulted in damage comparable to that caused by indiscriminate attacks upon civilians.[313]

The court also ruled that when military targets were concentrated in a comparatively small area, and where defense installations against air raids were very strong, that when the destruction of non-military objectives is small in proportion to the large military interests, or necessity, such destruction is lawful. Thus, in the judgment of the Court, because of the immense power of the atom bombs and the distance from Allied land forces, the atomic bombings of both Hiroshima and Nagasaki constituted "an illegal act of hostilities under international law as it existed at that time, as an indiscriminate bombardment of undefended cities."[314]

Nonetheless, the claimant as an individual was not entitled to claim damages on the plane of international law, nor was he able, as a

---

[312]International law as found in Hague Convention of 1907 IV - The Laws and Customs of War on Land,[5] and IX - Bombardment by Naval Forces in Time of War,[6] and the Hague Draft Rules of Air Warfare of 1922–1923.

[313]Falk, Richard. "The Shimoda Case: A Legal Appraisal of the Atomic Attacks Upon Hiroshima and Nagasaki" *American Journal of International Law,* 59: 4 (Oct., 1965), pp. 759-793.

[314]Mark Selden, 'A Forgotten Holocaust: U.S. Bombing Strategy, The Destruction of Japanese Cities, and American Way of War from the Pacific War to Iraq,' in Tanaka, Yuki and Young, Marilyn, eds. *Bombing Civilians: A Twentieth Century History,* New Press, 2009, pp.76 – 96.

result of the doctrine of sovereign immunity, to pursue a claim on the plane of municipal law. In these circumstances, the plaintiffs had no rights to lose as a result of the waiver contained in Article 19 (a) of the Treaty of Peace with Japan. The Japanese court] found against the plaintiffs because even though the *Shimoda* court held that the United States had in fact violated international law, it also noted that Japan had waived the right of its nationals to recover against the United States because of the 1951 treaty. Without a waiver of all war crime claims that could have been brought by either side, Japan and the United States might have wrangled endlessly about liabilities arising out of the war.

In 2007, Japanese Prime Minister Abe Shinzō offered in print an unprecedented apology to China. It begins with an acknowledgment that Japan's indiscriminate bombing of civilians living in the Nationalist Chinese wartime capital of Chongqing beginning in 1938 violated international law. In his view that act allowed the United States to find justification for its own devastating incendiary raids on Japan's capital. The prime minister also admitted that, by not capitulating to the United States once defeat became inevitable, the Japanese government essentially permitted the fire-bombing of Tokyo and thereafter the rest of urban Japan in 1945. That air raid of March 10, 1945, that destroyed much of Tokyo's built-up area, killed an estimated 100,000 people, injured at least 40,000, and displaced one million. To show its sincerity, the Japanese government agreed to provide financial compensation to survivors and bereaved family members, conduct a comprehensive survey of the dead, and build a memorial both to honor them and to serve as a reminder that the air raids had occurred. It offered compensation of 1.23 billion yen ($15 million). In March 2007, sixty-two years after the catastrophic Great Tokyo Air Raid, 112 survivors and surviving family members announced their intent to sue the government for redress. The following month, the plaintiffs, the oldest of whom was eighty-six and whose average age was seventy-four, filed the suit with the Tokyo District Court.

From May 2007 to May 2009, the plaintiffs presented a number of arguments to justify the demand for compensation and an apology. Arguing that he results of terror bombing during World War II were not uniformly or randomly scattered across the whole land and society, the plaintiffs presented casualty estimates, damage reports, and the personal experiences of a few of the million people. They charged that the state acted irresponsibly by failing to conduct a thorough survey of air raid casualties and to confirm the identities of the dead. A core argument is the contention that the United States, via its indiscriminate bombing of civilians in the March 10 raid, was liable to pay compensation.[315] Therefore, the Japanese government should be held responsible for agreeing to article 19(a) of the 1951 San Francisco Peace Treaty in which the country waived "all claims of Japan and its nationals against the Allied Powers and their nationals arising out of the war or out of actions taken because of the existence of a state of war…." Finally, plaintiffs argued that the defendant Japanese government violated numerous legal statutes, including Article 14 of the current constitution that guarantees equality under the law, by not providing financial relief to air raid sufferers while offering generous benefits to veterans and their families.[316]

In suing the Japanese government, the plaintiffs confronted a multi-layered defense bulwark that has protected the state against an onslaught of war-related redress claims over the last few decades. One layer of defense meant to deter compensation claims brought by Japanese citizens is the "endurance doctrine" (*jūninron*), which holds that, since Japanese civilians equally experienced severe hardship during a time of national emergency, no particular group can receive special treatment from the government. This endurance doctrine has usually withstood significant challenges, and when the state has been forced to provide assistance to civilians for war-related suffering – such as atomic bomb sufferers – it has acted to present the suffering as exceptional so as not to open itself up to further compensation claims.[317]

For its part, the Japanese government contended that since neither international customary nor treaty law prohibited the use of atomic bombs

[315]According to Article Three of the 1907 Fourth Hague Convention - Laws and Customs of War on Land.

[316]B. Fisher, B. "Japan's Postwar Compensation Litigation." *Whittier Law Review*, 22 (2000), pp. 35-65.

[317]Karacas, Cary, "Fire Bombings and Forgotten Civilians: The Lawsuit Seeking Compensation for Victims of the Tokyo Air Raids" *Asia-Pacific Journal*, November-December 2019.

at the time, the question of a violation of positive international law did not arise. The Court agreed that the use of nuclear weapons was not expressly prohibited by international law, but it felt that the use a particular weapon was also to be ascertained in light of the principles of international law applicable to the conduct of warfare, in particular the prohibition on indiscriminate bombardment of an undefended city and the prohibition on inflicting unnecessary suffering. Although no generally applicable treaty relative to aerial bombardment was in force at time of the bombing, the Court held, citing the basis of the Draft Rules of Air Warfare (1923)[318] and, by analogy, on the rules applicable to bombardment by land and naval forces, that the indiscriminate bombing of undefended cities was unlawful under customary law. It further stated that the principle of distinction between military and non-military objectives had not been nullified by the supposed adoption by the belligerents of a doctrine of total war. The bombings of Nagasaki and Hiroshima, as undefended cities, were thus held to have been unlawful acts.

While allowing the case to continue, and plaintiffs to present all arguments they could conceive, in December 2009, the Tokyo District Court ruled against the plaintiffs and ordered them to pay all costs associated with the trial. It ruled that it is not possible to determine conclusively that Article Three of the 1907 Fourth Hague Convention became a part of international customary law or that it gives individuals the right to seek damages when the laws of war are violated. Essentially adopting the state's long-held "endurance obligation" position, the court stated that it would set a discriminatory precedent by ruling in favor of the plaintiffs due to the fact that the almost the entire Japanese nation had suffered during the war.[319]

In another echo of previous court rulings on war-related compensation cases, the Tokyo District Court held that only a legislative solution could resolve the issue of civilians, and in particular air raid victims and orphans, who might warrant financial relief for wartime

---

[318]The Hague Rules of Air Warfare, The Hague, December, 1922-1923.

[319]International Red Cross. "Shimoda Case: Compensation Claim against Japan Brought by the Residents of Hiroshima & Nagasaki." https://ihl-databases.icrc.org/applic/ihl/ihl-nat.nsf/0/aa559087dbcf1af5c1256a1c0029f14d

losses. Individuals do not have rights under international law unless this was specifically recognized in a treaty, the Japanese Court ruled that there was no general way open to an individual to claim damages directly under international law. As far as the judgment on the issues of the waiver in Article 19(a) of the Peace Treaty with Japan and the plaintiffs' claim for compensation are concerned, the Court took to the view that individuals have no rights under international law unless specifically recognized in a treaty and that thus there was no general way open for individuals to claim damages directly under international law. Also, because of the doctrine of sovereign immunity, individuals are also not able to pursue a claim on the municipal law plane. Therefore, the plaintiffs did not lose any rights as a result of the waiver of claims by Japan under the Treaty of Peace with Japan of 1951. A claim for damages caused by a State to a national of another State could be based on diplomatic protection, but as it is widely recognized in classical international law, any such claim is in fact the State's own claim for damages suffered by its nationals and not the claim of an individual. Thus Japan could waive, and did waive, all its claims, including those deriving from diplomatic protection, against the US under the peace treaty of 1951. The Court further held that claims by Japanese nationals under domestic law had also been waived.[320]

Compensation for Enslavement of POWs

According to the Japanese military's own record, nearly 25% of 140,000 Allied Prisoners of War died while interned in Japanese prison camps where they were forced to work to assist the Japanese war effort, in clear violation of prevailing law of nations. As we have previously seen, American POWs died at a rate of 27%. More than 100,000 civilians and POWs died in the construction of the Burma Railway alone.

One of the principal arguments of the court in the *Shimoda* case, that the waiver of claims in the San Francisco peace treaty precluded any actions for damages by Japanese citizens against the U.S. government, was also used in the US legal system. In the case of *Mitsubishi Materials*

[320]Kōji, Ikeya. *Abandoned Air Raid Victims and the Absence of Postwar Compensation.* Tokyo: Kurieitibu Nijūichi, 2010.

Corporation et al. v. Frank H. Dillman et al., [321]dealing with a lawsuit by a former US prisoner of war against Mitsubishi for its part in the forced labor that he performed during the Second World War, the Superior Court of Orange County rejected the motion and referred to the Shimoda case as follows:

> [The Japanese court] found against the plaintiffs because even though the Shimoda court was willing to say that the United States had in fact violated international law, it also said that Japan had waived the right of its nationals to recover against the United States because of the 1951 treaty....

Without a waiver of all war crime claims that could have been brought by either side, Japan and the United States might have wrangled endlessly about liabilities arising out of the war. The California court wrote,

> Before us are claims by surviving American prisoners of war against a number of Japanese companies for whom they were forced to do slave labor during World War II. It is a remarkable case, one in which the Attorney General of the United States and the Attorney General of the State of California are on opposite sides. The immediate cause of the litigation is a relatively recent change to our state law which was intended to allow "Second World War slave labor" victims to bring a lawsuit to recover compensation under state law for their labor against private companies who benefited by that labor during the war.[322] Make no mistake about it -- this legislation actually creates new claims that would not otherwise exist. [323]

The California Supreme Court granted review on April 30, 2003 of its legislation regarding compensation of POWs who were forcibly recruited into Japan's war effort in violation of international law. While the case was pending before the California State Supreme Court, the

---

[321]https://www2.gwu.edu/~memory/data/judicial/POWs_and_Forced_Labor_US/MitsubishiMaterials/nov052003decision.pdf

[322] Subdivision (b) of Code of Civil Procedure section 354.6 provides in pertinent part: "Any Second World War slave labor victim . . . may bring an action to recover compensation for labor performed as a Second World War slave labor victim . . . from any entity or successor in interest thereof, for whom that labor was performed, either directly or through a subsidiary or affiliate."

[323]California Super. Ct. Nos. 814430 & 814594.

United States Supreme Court handed down its opinion in the *Garamendi* case.[324] That ruling invalidated California's Holocaust Victim Insurance Relief Act, holding that it was unconstitutional because it conflicted with express federal policy concerning the recovery of insurance proceeds confiscated by Nazi Germany. In light of the *Garamendi* ruling, the California Supreme Court transferred this case back to this court, with instructions to vacate its earlier decision and reconsider the matter in light of the federal court's *Garamendi* opinion.

In the meanwhile, a federal court of the Ninth Circuit had ruled regarding the very law at issue before the California courts.[325] The issue was law that allowed former American POWs who had been forced to do slave labor for Japanese companies during World War II to sue those companies. The Ninth Circuit held that such state law is unconstitutional because it intrudes on the federal government's power to make and resolve war claims.

The case of *Deutsch v. Turner* concerned the alleged wrongs committed by German rather than Japanese corporations. It was a Holocaust case at issue here, and it, unlike most of the Japanese cases, involves only a single plaintiff. Deutsch, a Jew, was born and raised in Hungary. Deutsch asserted that, in 1944, when he was a child, the Nazis took over his town and transported him and his brother Georg to Auschwitz. There the brothers were tortured and forced to work as slaves for 14-hour days, seven days a week. Their work was for the benefit of private corporations, which entered into agreements with the Nazi government, whereby they paid the Nazis less than the prevailing wage for the work of the slaves. In reviewing all the cases involving slave labor the Court wrote,

> These cases concern the terrible abuses that German and Japanese corporate interests inflicted both on civilians and on soldiers captured by German and Japanese military forces during the Second World War. The corporations and their managers, with the cooperation and encouragement of their governments, subjected many individuals to vicious cruelties

[324] *American Ins. Ass'n v. Garamendi* (2003) 539 U.S. 396, 123 S.Ct. 2374, 156 L.Ed.2d 376.

[325] *Deutsch v. Turner Corp.* (9th Cir.2003) 324 F.3d 692.

> and forced them to work long hours without pay. The slave workers were often underfed, physically beaten, exposed to dangerous conditions, and denied medical care; many were murdered, and others died as a result of the maltreatment they suffered.[326]

The federal district court dismissed the action as presenting a non-justiciable political question.[327] On appeal the Ninth Circuit held that the California Code of Civil Procedure section 354.6 is an unconstitutional intrusion on the foreign affairs power of the United States and that Appellants' remaining claims are barred by the applicable statutes of limitations. The judgments of the district courts were affirmed.

Compensation for Enslavement of Conquered Peoples

The extent of Japanese enslavement of conquered peoples is nearly beyond imagination. One rather common estimate projected that more than 10 million Chinese civilians were mobilized into forced labor. The U.S. Library of Congress estimates that in Java, between 4 and 10 million *romusha* ("manual laborer"), were forced to work by the Japanese military. Some 270,000 of these Javanese laborers were sent to other Japanese-held areas in South East Asia. Only 52,000 were repatriated to Java, suggesting that there was a death rate of approximately 80%. Korean authorities estimated that approximately 670,000 Koreans, were conscripted into labor from 1944 to 1945. Approximately 670,000 of them were taken to Japan, where about 60,000 died between 1939 and 1945 due in large to exhaustion, maltreatment, executions, undernourishment, and poor working conditions. The total deaths of Korean forced laborers in Korea and Manchuria for those years is estimated to be between 270,000 and 810,000.[328]

Imperial Japan had annexed the Empire of Korea in 1910 as an important first step in its colonization of most of Asia. Over a million

---

[326] *Deutsch v Turner,* United States Court of Appeals, Ninth Circuit, January 21, 2003, 324 F.3d 692 (9th Cir. 2003).

[327] *Deutsch v. Turner,* 2000 WL 33957691 (C.D.Cal. 2000).

[328] Rummel, R. J. *Statistics of Democide: Genocide and Mass Murder Since 1990.* Lit Verlag, 1999.

Koreans would be conscripted as slave labor for Japan's war effort. Additionally, there were hundreds of thousands of women forced into sexual slavery for the Japanese army. Those women were sent to military-run brothels. As many as 7.8 million Koreans were conscripted as forced labor or soldiers during Japan's imperial expansion before and during World War II, according to South Korean estimates. They toiled in mines and munitions factories across Asia and, in some cases, forced to fight alongside Japanese troops.

In 1965, Japan and Korea entered into a treaty arrangement that established basic diplomatic relationship between the two countries in the aftermath of imperial Japan's defeat in World War II and Korea's regaining its independence. For its part, Japan has long argued that the 1965 treaties put an end to any obligation for Japan to compensate South Korea for its colonialism and war crimes, as Japan paid $300 million (the equivalent of about $2.28 billion today) and loaned South Korea an additional $500 million to settle Korea's claims.

Despite the treat arrangement in 2005, four former Korean forced laborers sued Nippon Steel & Sumitomo Metal Corporation, the successor of the wartime steel mill, in the South Korean courts after losing an earlier lawsuit they filed in Japan. Following thirteen years of litigation, in October 2018, the plaintiffs prevailed in the Supreme Court of Korea.

Rather than accepting the Korean court's judgment, Japan accused Korea of "trying to shift South Korea's responsibility" over the wartime forced labor, ruling as though Korea had been the aggressor. Following the Korea Supreme Court's decision, Tokyo has threatened to recall its ambassador to South Korea, levy sanctions against South Korean exports to Japan, seize South Korean government property in Japan, and reintroduce visa requirements for visiting Korean tourists. The Japanese government even demanded that the South Korean President invalidate the Supreme Court's decision. Japan claims its immunity stems from respect for international law, and that South Korea has violated the treaties into which the two countries entered in 1965.

The claim of immunity based upon the legitimacy of the 1965 treaties is more tendentious than Japan is willing to admit. In examining Japan's legal stance, it is far from clear whether the 1965 treaties actually waived private claims held by individual Koreans who were victimized by imperial Japan. The diplomatic documents related to the negotiation process of the 1965 treaties were not made public until 2005. These documents show that Japan consistently refused to acknowledge it had any liability to Korea arising from its colonial rule, as Japan refused to acknowledge the illegitimacy of its colonization of Korea. Accordingly, the 1965 treaties make no reference to whether Japan's colonial rule over Korea had any legitimacy.[329]

Japan's negotiators for the 1965 treaties had argued that Japan owed nothing to Korea. The text of the treaty makes no reference as to whether imperial Japan committed any wrong to Korea for which compensation. Whatever money Japan paid to Korea pursuant to the treaties settled nothing because Japan admitted nothing. Indeed, the Japanese government in 1965 characterized the payment not as reparations for a wrong, but as a congratulatory gesture for Korea's independence. Under international law, as well as logically, a state cannot bargain in a treaty to settle something it had never admitted to in the first place.

Under international law, even if Japan had intended to make a bargain of this kind, such agreement would be void based on well-established principles of international law. *Jus cogens*, or peremptory norm, provides that a state cannot sign away its people's human rights through a treaty. Article 53 of the Vienna Convention on the Law of Treaties states: "A treaty is void if, at the time of its conclusion, it conflicts with a peremptory norm of general international law."

It is beyond dispute that the Japanese Empire practiced wartime slavery against the Korean people. Slavery violates basic norms of the law of nations. Indeed, the prohibition of slavery is one of the earliest peremptory norms to be recognized in modern international law, along with prohibitions against torture and genocide of civilians. The validity of a peremptory norm does not depend on the perpetrator's understanding of it as a violation at the time. Still more damning, it is also absolutely clear that imperial Japan recognized prohibition of slavery as an international peremptory norm, as it was a signatory to treaties prohibiting slavery as

---

[329]*New York Times,* 13 February 2019. See also http://www.xinhuanet.com/ english/ 2019-08/14/c_138308682.htm

early as 1925. Even if the South Korean government intended to waive individual Koreans' claims arising from imperial Japan's conscription for slave labor, such attempts would have no effect under the prevailing law of nations.[330]

[330]https://foreignpolicy.com/2019/05/29/tokyo-keeps-defending-world-war-ii-atrocities/

Bibliography

Asada, Sadao. *From Mahan to Pearl Harbor*. Naval Institute Press, 2006.

Barenblatt, Daniel. *A Plague Upon Humanity: The Secret Genocide of Axis Japan's Warfare Operation.* Harper Collins, 2005.

Boister, N. & R. Cryer. *The Tokyo International Military Tribunal: A Reappraisal.* Oxford University Press, 2008.

Borch, Frederic L.; Martinez, Daniel, Kimmel, *Short, and Pearl Harbor: The Final Report Revealed,* Naval Institute Press, 2005.

Bryan, Joseph, and Philip Reed. *Mission Beyond Darkness.* Duell, Sloan and Pearce, 1945.

Buell, Thomas B. *The Quiet Warrior: A Biography of Admiral Raymond A. Spruance.* 1987.

Chang, Iris. *Rape of Nanking*. London: Penguin, 1998.

Coggins, Jack, *The Campaign for Guadalcanal: A Battle That Made History.* Garden City, NY: Doubleday and Co., 1972.

Como, Byron. *The Defenders of Taffy 3*. Seattle: Amazon, 2017.

Como, Bryan. *The Raiders of Taffy 3*. Seattle: Amazon, 2017

Cowl, Philip A. "Alfred Thayer Mahan: The Naval Historian," in Paret, Peter, Gordon A. Craig, and Felix Gilbert, eds. *Makers of Modern Strategy from Machiavelli to the Nuclear Age* (1986).

Cox, Robert Jon. *The Battle Off Samar: Taffy III at Leyte Gulf.* Agogeebic Press; 5th ed., 2010.

Cunliffe, William H. "Nazi War Crimes and Japanese Imperial Government Records: Select Documents on Japanese War Crimes and Japanese Biological Warfare, 1934-2006." Interagency Working Group.

Cutler, Thomas. *The Battle of Leyte Gulf: 23–26 October 1944*. Annapolis, Maryland, U.S.: Naval Institute Press, 1994.

D'Albas, Andrieu. *Death of a Navy: Japanese Naval Action in World War II.* Devin-Adair, 1965.

Daws, Gavan, *Prisoners of the Japanese*. Morrow, 1994.

Dorpalen, Andreas. *The World of General Haushofer: Geopolitics in Action.* New York, Farrar and Rinehart, 1942.

Dower, John, *War Without Mercy: Race and Power in the Pacific War.* New York: Pantheon, 1985.

Drea, Edward; Bradsher, Greg; Hanyok, Robert; Lide, James; Petersen, Michael, & Daqing Yang. *Researching Japanese War Crimes: Introductory Essays.* Washington, D.C.: Nazi War Crimes and Japanese Imperial Government Records Interagency Working Group, 2006.

Dull, Paul S. *A Battle History of the Imperial Japanese Navy, 1941–1945*. Naval Institute Press, 1978.

Escoda, Jose M., *Warsaw of Asia: The Rape of Manila*, revised ed. Quezon City, Philippines: Giraffe Books, 2001.

Falk, Stanley, *Bataan: The March of Death.* New York: W. W. Norton, 1962.

Feldhaus, J. Gordon, "The Trial of Yamashita," *South Dakota Bar Journal*, 15: 181-93 [1946].

Felton, Mark. *The Devil's Doctors: Japanese Human Experiments on Allied Prisoners of War.* Pen & Sword, 2012.

Field, James A. *The Japanese at Leyte Gulf: The Sho operation.* Princeton University Press, 1947.

Friedman, Kenneth. *Afternoon of the Rising Sun: The Battle of Leyte Gulf.* Presidio Press, 2001.

Fuchida, Mitsuo and Masatake Okumiya. *Midway: The Battle that Doomed Japan, the Japanese Navy's Story.* Annapolis, Maryland: Naval Institute Press, 1955.

Fuchida, Mitsuo and Okumiya, MasatakeGold, Hal. *Unit 731 Testimony: Japan's Wartime Human Experimentation Program.* Tuttle, 2004.

Ganoe, John, "The Yamashita Case and the Constitution" *Oregon Law Review*, April 1946, 143-58.

Gilmore, Allison B. *You Can't Fight Tanks with Bayonets: Psychological Warfare against the Japanese Army in the South West Pacific.* Lincoln, Nebraska: University of Nebraska Press, 1998.

Gold, Hal. *Unit 731 Testimony: Japan's Wartime Human Experimentation Program.* Tuttle, 2004.

Goldstein, Donald M. *Midway: The Battle that Doomed Japan*. Hutchinson, 1957.

Goldstein, Donald M. and Dillon, Katherine V., eds., *The Pearl Harbor Papers: Inside the Japanese Plans*, Brassey's, 2000.

Handelman, Stephen and Alibek, Ken. *Bio-hazard: The Chilling True Story of the Largest Covert Biological Weapons Program in the World—Told from Inside by the Man Who Ran It*, Random House, 1999.

Harris, Sheldon H. *Factories of Death: Japanese Biological Warfare, 1932-45 and the American Cover-Up*. Routledge, 2004.

Hastings, Max, *Retribution: The Battle for Japan, 1944-45*. Knopf, 2008

Hattendorf, John B., ed. *The Influence of History on Mahan.* Naval War College Press, 1991

Haushofer, Karl. *Der Kontinentalblock : Mitteleuropa, Eurasien, Japan,* Berlin : Eher, 1941.

Havers, R. P. W., *Reassessing the Japanese Prisoner of War Experience: The Changi POW Camp, Singapore, 1942–45.* London: Routledge Curzon, 2003

Haushofer, Karl. *Das Japanische Reich in seiner geographischen Entwicklung.* Vienna: L.W. Seidel & sohn, 1921.

Hersey, John, *Into the Valley: Marines at Guadalcanal.* Lincoln: University of Nebraska Press, 2002.

Heske, Henning, "Karl Haushofer: His Role in German Politics and in Nazi

Politics," *Political Geography* 6 (1987), pp. 135–144.

Hicks, George, *Comfort Women: Japan's Brutal Regime of Enforced Prostitution in the Second World War.* New York: Norton, 1997.

Holmes, W., *Double-Edged Secrets: U.S. Naval Intelligence Operations in the Pacific During World War II.* Annapolis: Naval Institute Press, 1979.

Honeycutt, Fred L. *Military Rifles of Japan*, Julin Books, 1996.

Hornfischer, James D. *The Last Stand of the Tin Can Sailors.* Bantam, 2004

Hoyt, Edwin P., *Pearl Harbor,* G. K. Hall, 2000.

Isom, Dallas Woodbury. *Midway Inquest: Why the Japanese Lost the Battle of Midway.* Bloomington, Indiana: Indiana University Press, 2007

Jacobsen, Hans-Adolf. Karl Haushofer: Leben und Werk. 2 vols. Harald Boldt Verlag, Boppard 1979.

Japanese Prisoners of War, eds. Philip Towle et al. London: Hambledon and London, 2000.

Kernan, Alvin, *The Unknown Battle of Midway*. New Haven, Connecticut: Yale University Press, 2005.

Khabarovsk War Crime Trials. Materials on the Trial of Former Servicemen of the Japanese Army Charged with Manufacturing and Employing Biological Weapons, Moscow: Foreign Languages Publishing House, 1950.

Knox, Donald, *Death March: The Survivors of Bataan.* New York: Harcourt Brace Jovanovich, 1981.

Kratoska, Paul, ed., *Southeast Asian Minorities in the Wartime Japanese Empire.* London: Routledge Curzon, 2002.

Kratoska, Paul H. *Asian Labor in the Wartime Japanese Empire.* Aemonk, NY: Sharppe, 2005

Lael, Richard L., *The Yamashita Precedent.* Wilmington, DE: Scholarly Resources, 1982.

Landas, Marc, *The Fallen: A True Story of American POWs and Japanese Wartime Atrocities.* Wiley, 2004.

Leckie, Robert, *Challenge for the Pacific: The Bloody Six-Month Battle Of Guadalcanal.* New York: Da Capo Press, 1999.

Li, Peter, *Japanese War Crimes*. Transaction Publications, 2013.

Lundstrom, John. *The First Team*. U. S. Naval Institute Press, 1990.

Mackinder, Halford. *Democratic Ideals and Reality*, New York: Holt, 1919.

Mahan, Alfred Thayer. *The Interest of America in International Conditions* (1910). Reprint by Transaction Books, 2003.

Mahan, Alfred Thayer, *The Problem of Asia and Its Effect Upon International Policies* (1905) Reprint by The Classics, 2013.

Materials on the Trial of Former Servicemen of the Japanese Army, Charged with Manufacturing and Employing Bacteriological Weapons. Moscow: Foreign Languages Publishing House, 1950.

McCollum, Duncan O. *Japanese Rifles of World War II*. Excalibur, 1996.

McLynn, Frank. *The Burma Campaign: Disaster into Triumph, 1942–45.* Yale University Press; 2011.

Minear, Richard R., *Victors' Justice: The Tokyo War Crimes Trials*. University of

Michigan Press, 1971

Morison, Samuel Eliot, *New Guinea and the Marianas, March 1944–August 1944,* in *History of United States Naval Operations in World War II,* 1950.

Nozaki, Yoshiko, *Textbook Controversy and the Production of Public Truth: Japanese Education, Nationalism, and Saburo Ienaga's Court Challenges. University* of Wisconsin—Madison, 2000

Peattie, Mark R. *Sunburst: The Rise of Japanese Naval Air Power, 1909–1941.* Annapolis, Maryland: Naval Institute Press, 2013.

Peattie, Mark and David Evans, *Kaigun.* U.S. Naval Institute Press, 1997.

Peattie, Mark R. and Evans, David C., *Kaigun: Strategy, Tactics, and Technology in the Imperial Japanese Navy,* Naval Institute Press, 1997.

Piccagallo, Philip R. *The Japanese on Trial: Allied War Crimes Operations in the East, 1945-1951.* University of Texas Press, 1980.

Poor, Henry V., Henry A. Mustin and Colin G. Jameson. *The Battles of Cape Esperance, 11 October 1942 and Santa Cruz Islands, October 26, 1942.* Washington, DC: Naval Historical Center, 1994.

Potter, E. B. *Bull Halsey.* Naval Institute Press, 2003.

Prange, Gordon, *Miracle at Midway.* McGraw-Hill, 1982

Prange, Gordon William; Goldstein, Donald M. and Dillon, Katherine V. *December 7, 1941: The Day the Japanese Attacked Pearl Harbor.* McGraw-Hill, 1988.

Roosevelt, Theodore, *The Naval War of 1812* (1882). Reprint by Modern Library, 1999.

Rummel, R.J., *Genocide and Mass Murder Since 1900.* Charlottesville, VA: School of Law, University of Virginia, 1997

Russell. Lord Russell of Liverpool, *The Knights of Bushido: A Short History of Japanese War Crimes.* Berkeley Books, 1961.

Schmidt, David Andrewt, *Ianfu–The Comfort Women of the Japanese Imperial Army of the Pacific War: Broken Silence.* Lewiston, NY: Edwin Mellen Press, 2000

Spang, Christian W., *Karl Haushofer und Japan. Die Rezeption seiner geopolitischen Theorien in der deutschen und japanischen Politik,* Munich: Iudicium, 2013

Seager II, Robert, ed. *Letters and Papers of Alfred Thayer Mahan.* 3 vols. 1975

Seager, Robert, *Alfred Thayer Mahan: The Man and his Letters.* Annapolis: Naval Institute Press, 1977

Sempa, Francis P. *Geopolitics: From the Cold War to the 21st Century.* Transaction Books, 2002

Sides, Hampton "The Trial of General Homma," American Heritage, February/March 2007.

Smith, Carl, *Pearl Harbor 1941: The Day of Infamy*; Osprey Campaign Series #62, Osprey Publishing, 1999.

Smith, Douglas V. *Carrier Battles: Command Decision in Harm's Way.* Annapolis, Maryland: Naval Institute Press, 2006.

Smith, Peter C. *Midway Dauntless Victory; Fresh perspectives on America's*

Seminal Naval Victory of 1942. Barnsley, UK: Pen & Sword, 2007.

Stewart, Adrian. *The Battle of Leyte Gulf.* Hale, 1979.

Stille, Mark E., *Tora! Tora! Tora!: Pearl Harbor 1941*; Osprey Raid Series #26, Osprey Publishing, 2011

Stille, Mark. *USN Carriers vs IJN Carriers: The Pacific 1942*. New York: Osprey, 2007.

Tagaya, Osamu: "The Imperial Japanese Air Forces," in: Higham, Robin & Harris, Stephen, eds. *Why Air Forces Fail*. University of Kentucky, 2006.

Tanaka, Yuki, *Hidden Horrors: Japanese War Crimes in World II*. Boulder: Westview Press, 1996.

Tanaka, Yuki. *Japan's Comfort Women: Sexual Slavery and Prostitution During World War II and the U.S. Occupation. London: Routledge, 2001*

Tillman, Barrett, *Clash of the Carriers: The True Story of the Marianas Turkey Shoot of World War II.* Penguin, 2006.

Tsuchiya, Takashi, "The Imperial Japanese Experiments in China" in *Oxford Textbook of Clinical Research Ethics*, 2008.

Tsuji, Masanobu, *Japan's Greatest Victory, Britain's Worst Defeat*. Trans. by Margaret E. Lake. Da Capo Press, 1997.

Turk, Richard W. *The Ambiguous Relationship: Theodore Roosevelt and Alfred Thayer Mahan.* Greenwood Press, 1987.

Tsuchiya, Takashi, *"The Imperial Japanese Experiments in China"* in *Oxford Textbook of Clinical Research Ethics, 2008.*

Twining, Merrill B. *No Bended Knee: The Battle for Guadalcanal*. Novato, CA.: Presidio, 1996.

Wakabayashi, Bob Tadashi, "The Nanking 100-Man Killing Contest Debate: War Guilt Amid Fabricated Illusions," *Journal of Japanese Studies* 26: 2, [Summer 2000)]: 307–4

Willmott, H. P. *The Battle of Leyte Gulf: The Last Fleet Action*. Indiana University Press, 2005.

Wimmel, Kenneth, *Theodore Roosevelt and the Great White Fleet: American Seapower Comes of Age*. Washington: Brassey's, 1988.

Woodward, C. Vann. *The Battle for Leyte Gulf.* (1947) Skyhorse Reprint, 2007.

Yamada, Otozo. *Materials on the Trial of Former Servicemen of the Japanese Army, Charged with Manufacturing and Employing Bacteriological Weapons*. Moscow: Foreign Languages Pub. House, 1950.

Yamamoto, Masahiro, *Nanking: Anatomy of an Atrocity.* Westport, CT: Praeger, 2000.

Yang, Daqing, "Convergence or Divergence? Recent Historical Writings on the Rape of Nanking" 104 *American Historical Rev.* (June 1999) 842-65.

Y'Blood, William T. *Red Sun Setting: The Battle of the Philippine Sea*. Annapolis, Maryland: US Naval Institute Press, 1981.

Yoshihara, Toshi and Holmes, James, *Chinese Naval Strategy in the 21st Century: The Turn to Mahan.* Taylor & Francis, 2009.

Zimm, Alan D., *Attack on Pearl Harbor: Strategy, Combat, Myths, Deceptions*, Havertown, PA: Casemate, 2013

Index

B

C